# Health Technical Memorandum 2011

### Design considerations

## Emergency electrical services

London : HMSO

**NHS** Estates

An Executive Agency of the Department of Health

**HMSO**
**Standing order service**

Placing a standing order with HMSO BOOKS enables a
customer to receive future titles in this series automatically
as published. This saves the time, trouble and expense of
placing individual orders and avoids the problem of
knowing when to do so. For details please write to HMSO
BOOKS (PC 13A/1), Publications Centre, PO Box 276,
London SW8 5DT quoting reference 14 02 017. The
standing order service also enables customers to receive
automatically as published all material of their choice which
additionally saves extensive catalogue research. The scope
and selectivity of the service has been extended by new
techniques, and there are more than 3,500 classifications to
choose from. A special leaflet describing the service in detail
may be obtained on request.

# About this publication

Health Technical Memoranda (HTM) give comprehensive advice and guidance on the design, installation and operation of specialised building and engineering technology used in the delivery of health care.

They are applicable to new and existing sites, and are for use at various stages during the inception, design, construction, refurbishment and maintenance of a building.

**Health Technical Memorandum 2011**
HTM 2011 focuses on the:

- legal and mandatory requirements;

- design applications;

- maintenance; and

- operation,

of emergency and essential electrical supply equipment in all types of health care and personal social services premises.

It is published as four separate volumes each addressing a specialist discipline:

- This volume – **Design considerations** – highlights the overall requirements and considerations that should be applied to the design up to the contract document;

- **Management policy** – outlines the overall responsibility of managers of health care and personal social services premises, and details their legal and mandatory obligations in setting up and operating a reliable emergency electrical supply. It summarises the technical aspects involved and concludes with a list of definitions;

- **Validation and verification** – details the requirements for ensuring that manufactured equipment is formally tested and certified as to contract and manufactured to the highest level of quality assurance. The importance of commissioning is emphasised and the order of tests on site is listed. Routine testing, which is a subset of these commissioning tests, is also reviewed.

- **Operational management** – provides information for those responsible for overseeing and operating day-to-day running and maintenance. Coverage includes stand-by generators, batteries, test runs and record keeping. Supplementary power for peak periods, and combined heat and power systems are also discussed.

Guidance in this Health Technical Memorandum is complemented by the library of National Health Service Model Engineering Specifications. Users of the guidance are advised to refer to the relevant specifications for "Emergency supply".

The contents of this Health Technical Memorandum in terms of management policy, operational policy and technical guidance are endorsed by:

- The Welsh Office for the NHS in Wales;

- The Health and Personal Social Services Management Executive in Northern Ireland;

- The National Health Service in Scotland Management Executive,

and set the required standards consistent with Departmental Cost Allowances.

References to legislation appearing in the main text of this guidance apply in England and Wales. Where references differ for Scotland and/or Northern Ireland these are given as marginal notes.

Where appropriate, marginal notes are also used to amplify the text.

# Contents

# 1.0 Scope

## General

**1.1** Health care and personal social services premises are totally dependent upon electrical power supplies, not only to maintain a safer and more comfortable environment for patients and staff, but also to give greater scope for treatment using sophisticated medical equipment at all levels of clinical and surgical care. Changes in application, design and statutory requirements have led to the introduction of a new generation of equipment and new standards of reliability.

**1.2** Interruptions in electrical power supplies to equipment can seriously disrupt the delivery of health care with serious consequences for patient well-being. Health care and personal social services premises must therefore ensure that they can continue to provide electrical power to essential services in the event of prolonged or short disruption to supplies.

## Emergency electrical services

**1.3** Emergency electrical services form an integral part of the health care and personal social services premises supply network in meeting both safety and functional requirements. They can be in the form of batteries, uninterrupted power supply (UPS) systems, or stand-by generators.

**1.4** The provision of emergency electrical services in health care and personal social services premises is a management responsibility at both new and existing sites. This guidance is equally applicable to premises which offer acute health care services under the Registered Homes Act 1984.

*Nursing Homes and Agencies Act (Northern Ireland) 1971*
*Nursing Homes Regulations (Northern Ireland) 1974*
*Registered Establishments (Scotland) Act 1987*

**1.5** This volume – **Design considerations** – comprises five main sections:

a. **General**

This is an introductory section, giving general design advice and requirements for normal and emergency electrical supplies and equipment. A guide to standards, relevant references and a set of electrical circuit diagrams is included.

b. **Lighting and power supplies**

This section outlines the requirements relating to emergency lighting, as recommended in BS5266, Part 1 'Code of Practice for Emergency Lighting in Premises'. Advice is given on the types of acceptable emergency luminaires and general illumination in medical and plant areas of the health care and personal social services premises. For more detailed design requirements, reference can be made to the Chartered Institution of Building Services Engineers (CIBSE) document 'Lighting Guide for Hospitals and Health Care Buildings'. Use of uninterruptible power supplies to provide a no break, interference-free electrical supply to specialised medical/surgical and electronic data equipment is discussed.

c. **Engines**

This section considers the various types of internal combustion engine available to drive an emergency generator in accordance with BS5514 (ISO 3046). A review of engine exhaust pollution, methods of exhaust gas cleaning and the stack height required is introduced to give a basic understanding of the need for environmental considerations where electrical generation may be expanded to the role of combined heat and power. The layout of engine rooms and basic facilities required to assist maintenance work is discussed.

### d. Generators

Here the overall design and control of AC generators is outlined together with the application of controls for single or parallel generation, when connected to or separate from the regional electricity company supply. Electrical operational problems are covered where AC generators or induction generators are run in parallel with the regional electricity company network. An inventory of control panel instrumentation for engine-generator units is also included.

### e. Batteries for emergency supply

Different types of battery are outlined together with their individual advantages and disadvantages for emergency use.

# 2.0 General

## Normal electrical supplies

**2.1**   The regional electricity companies have in the past regarded hospitals and other health care and personal social services premises as priority users. This arrangement cannot be fully guaranteed in the future. Failures will inevitably occur and experience indicates that the risk of failure is higher in rural districts, where there is wider use of overhead line distribution. The regional electricity company should be consulted to assess the extent to which the normal supply may be at risk from interruptions from various causes.

**2.2**   Large premises with over 300kW demand should be supplied by a high voltage feed (for example, 11kV) at three-phase to a sub-station located within the health care and personal social services premises. Smaller premises are normally supplied by a 415V, three-phase feed. High voltage supplies are more reliable.

**2.3**   In the Nucleus hospital design, additional security at the point of 11kV supply may be introduced by the use of two transformers of equal rating. One transformer supplies the non-essential 415V, three-phase switchboard and the other the essential 415V, three-phase switchboard. The emergency generators are connected to the essential switchboard to auto-start and load within 15 seconds on loss of mains supply. The outputs of both transformers are connected by fully rated circuit breakers and protected to their respective switchboards. The switchboards are interconnected by a fully rated bus coupling circuit breaker. The switching logic between the three circuit breakers will be a "two closed, one open" arrangement to restrict the maximum prospective fault current (MPFC) at the switchboards to a connection of one transformer and all emergency generators installed. The electrical protection must be designed for this requirement. This arrangement can utilise the generators, when synchronised to the normal supply for peak lopping, or as purpose-built generating units operating in "island" mode.

**2.4**   Having two separate 11kV, three-phase supply feeders is an additional safeguard for larger premises, and whether this is practicable largely depends on the local distribution system. The regional electricity company will advise. Provision of the second supply will only be justified where it can be shown to be economically advantageous.

**2.5**   The electrical installation within the health care and personal social services premises should be designed to limit interruptions in the supply due to internal faults as far as reasonably practicable. This is covered in greater detail in Health Technical Memorandum 2007 – 'Electrical Services: Supply and Distribution' (in preparation).

## Emergency electricity supplies

**2.6**   Although failures in the normal electricity supply are likely to be of short duration, all emergency generating sets should be designed and rated to provide continuous full-load for prolonged periods for all essential functions.

**2.7** Generating plant must be available to provide electrical power to those areas which will enable the health care and personal social services premises to carry out their essential functions. Within this general objective, the aim should be to keep electrical installations as simple as practicable and avoid unnecessary segregation and duplication of essential and non-essential circuits. This applies particularly for lighting circuits, where comparatively small loads are involved.

**2.8** Where new health care and personal social services premises are built in separate phases the emergency power supply for the whole premises should, as far as possible, be planned and evaluated at the design stage. This will enable the total emergency power supply requirement to be assessed in the planning stages and appropriate areas of accommodation allocated. The required AC generator sets should be installed early in each phase to make the maximum amount of emergency power available as early as possible and so that staff training can be carried out.

**2.9** Emergency power supplies for existing health care and personal social services premises should be periodically reassessed and improved where necessary to ensure that they remain adequate to maintain essential clinical and surgical life-support facilities.

**2.10** It is recommended that managers make emergency arrangements for 415V, three-phase, 50Hz mobile generators of suitable rating to be obtainable elsewhere at short notice to supplement existing sources. Suitably rated switchgear should be identified within the 415V, three-phase essential services system of the health care and personal social services premises provided with graded protection and interlocked ready to be utilised for temporary emergency connection. An earth bar connection should be available for connection to the earth terminal of the generator base plate. Ideally this may be achieved by a "plug-in" facility rated to BS4343 (IEC 309): 'Industrial plugs, socket-outlets and connectors'.

**2.11** In planning an installation, it is desirable to reserve a parking space for mobile emergency generators where they may be easily connected to an allocated switch or "plug-in" point.

**2.12** A common difficulty where the emergency generator cannot be synchronised to the normal supply is in providing routine test loads to check engine performance. Considerable negotiation is usually necessary with medical and surgical departments to agree suitable times of least disruption and/or sufficient demand for load testing the generators. It should be noted that this test does not constitute a full operational test for automatic controls. The provision of TPN and E isolated terminals, located between the emergency generator(s) circuit breaker terminals and the auto changeover switch, will enable temporary connection to a ballast load bank to absorb all or part of the emergency generator output. The provision of a suitable auto-changeover switch with a bypass facility would also allow routine tests on the auto-changeover switch. Load bank ballast resistors and inductors can be obtained as fixed or mobile equipment with variable control and ratings up to 1.0MW. They may be purchased or hired from specialist manufacturers.

**2.13** A "plug-in" socket for a mobile ballast load bank may be made, as proposed for the mobile generating set.

**2.14** Parking space for this operation should be allocated adjacent to the location selected.

# Segregated or unified circuits in new installations

**2.15** The rapid increase in demand for health care and personal social services premises offering advanced surgical and medical care has led to a proportionately increased demand for essential services. When planning for new installations the option of segregated non-essential and essential electrical systems or a unified electrical system must be evaluated.

**2.16** The provision of two segregated systems, each of smaller power capacity, must be balanced against having one larger unified power system in terms of economics and reliability in emergencies.

**2.17** Even when two segregated systems are provided it is desirable to have an emergency coupling provision normally locked open between them. This allows the standby generator to be connected to both systems if necessary, for example during a prolonged outage some normally non-essential services may become "essential" such as catering and laundry. Also, with the coupling it is possible to provide a larger test load.

**2.18** For single running AC emergency generators required to supply only segregated essential services, a main auto-changeover switch should be provided. It should be connected to supply 415V, three-phase power to the health care and personal social services premises from two sources: either from the regional electricity company's normal 415V, three-phase supply via the main switchboard to the essential services switchboard, or, in an emergency, from the AC emergency generator to only the essential services switchboard (see Figure 1 on page 16). The main auto-changeover switch and connecting cables must be rated for the essential load current and the maximum prospective fault current of the normal supply at the main and essential switchboards. For "island" running, the cables connecting to the emergency generators need only be rated to the total emergency generator's load and short circuit currents.

**2.19** Parallel running AC generators, which supply the non-essential and essential services electrical loads and also contribute to peak lopping, are connected differently from "island" running purpose-built emergency generators. An auto-changeover switch is not required, and can be replaced by a circuit breaker, connecting the main and essential services switchboards.

**2.20** The AC generators should be connected either by circuit breakers directly to the essential services switchboard or by a tie feeder from a generator switchboard. A function of the main transformer normal supply circuit breaker would be to trip and initiate the engine starts on loss of normal supply. The generators would also require individual synchronisation facilities for operational flexibility. All circuit breakers and cables will require to be rated for the combined prospective fault current from the 415V normal supply and generators (see Figure 14 on page 29).

**2.21** Detection is required on each of the three phases of the normal supply. Any single-phase voltage failure in the normal supply must initiate a start-up of the AC emergency generator set. The engine should start up and run up to speed and the generator should excite to 415V at 50Hz, allowing its circuit breaker to close. An auto-changeover switch, contactor device or circuit breaker where provided, would then be initiated to change to the AC generator supply. This operation should have an overall response time not exceeding 15 seconds. On stable return of the normal supply, the switching process reverses and the AC generator is eventually shut down.

**2.22** Provision should be made using a suitable timing relay for a minimum run time of up to 20 minutes, depending on local circumstances. This feature will provide time for the restored supply to stabilise and for starter batteries to recharge.

**2.23** It is important that each principal phase voltage detector monitoring the presence of normal 415V, three-phase supply should be fuse protected at the bus bars and located in the incoming supply main switchboard. Loss of power supply by fuse operation in a downstream sub-circuit must not be mistaken for a complete failure in the normal supply from the regional electricity company via the main supply transformer.

**2.24** Where a single cable is used to supply both non-essential and essential services, the supply circuit should be arranged, where possible, to provide an alternative supply to safeguard the essential services.

**2.25** In large segregated installations it may be necessary to divide the essential services supply into two or more sub-main feeder circuits, depending upon location and load demand. The lengths of cable should be optimised for either rated current or a maximum volt-drop of 4%.

**2.26** Where a dual supply for an emergency service is required using generated or battery supplies, a discrete auto-changeover contactor with separate voltage sensing should be provided. Figure 1 on page 16 shows a typical arrangement.

**2.27** Auto-changeover switches in segregated sub-circuits should only be used in emergency areas. In other essential supply circuits, manual changeover switching between essential and non-essential supply circuits may be provided. These manual changeover switches should be clearly identified by approved orange/red markers.

**2.28** It is important to regulate any step loading of the AC generator and the driving engine to within the control range of the engine acceptance load category and generator rating. Time controlled switching should be used to graduate shock electrical loading. Dead loads should be kept to a minimum and reserved for only the most important of essential services.

**2.29** The AC emergency generator should always be connected to a main auto-changeover switch by a fully rated withdrawable circuit breaker that can be isolated for set maintenance and is provided with the correct discriminating electrical protection for the generator and circuits.

**2.30** An earth connection to the neutral star point of an emergency AC generator winding is required to provide an earth and neutral continuity connection for the 415V, three-phase supplies when not connected to the regional electricity company normal supply transformer. When AC generators are operated in parallel or in parallel with the regional electricity company supply, provision should be made for a neutral switchboard to ensure that only one earth connection can be selected – see paragraph 5.3.

**2.31** All motor starters, except where specifically required otherwise, should have delayed no-volt release main contactors. Local/remote/auto-control selection and timer delayed auto-starting should be provided where essential.

## Segregated or unified circuits in existing installations

**2.32** Where additional essential supply facilities are being provided for existing installations the choice of distribution will depend on local conditions and the location of the existing essential supply relative to the electrical load centre in the health care and personal social services premises.

**2.33**  Transfer of existing essential lighting or domestic small power circuits to new essential services distribution boards directly connected to the essential services switchboard should be considered.

**2.34**  In general, essential lighting circuits and radial or ring main domestic small power circuits may share a main distribution circuit. Each localised area should have a common distribution board with separate circuits. Lighting circuits are more suitable than power circuits for sub-circuit auto-changeover contactors, especially when connected to an emergency battery supply. Control of AC sub-circuit auto-changeover switches should be time graded for step-ramp loading of the engine driving the emergency generator.

**2.35**  Where only limited emergency facilities are required, such as lighting of escape routes, self-contained emergency luminaires are sufficient.

**2.36**  All existing voltage loss detector locations for lighting or power auto-changeover contactors should be reappraised to ensure they provide maximum safeguard against a main or sub-circuit spurious changeover resulting from in-circuit equipment faults.

## Selection of emergency supply equipment

**2.37**  Reliability in service is of prime importance for equipment used in essential supplies. Goods and services subject to the assessment and inspectorate services based on BS5750 (ISO 9000, EN 29000) 'Quality Systems', and BS9000 'Electronic Components of Assessed Quality' should be encouraged and supported. In particular, special attention should be given to the reliability of contactors, relays and solid state electronic devices and batteries that are used for starting up emergency plant and for changing over from normal to essential supplies. Experience has shown that many failures are attributable to these components. BS5992 'Electrical Relays', BS5424, Part 1, 1977, 'Contactors', and BS6667, Parts 1, 2, 3, 1985 or the IEC 801 standard for electromagnetic compatibility, refer. See also HTM 2014 – 'Abatement of electrical interference' (in preparation).

**2.38**  The likely maximum demand from essential loads should be carefully analysed to ensure that plant capacity is sufficient to supply the required loads in the event of a prolonged interruption to normal supplies.

**2.39**  Compliance of all electrical equipment with the appropriate British Standards or IEC, ISO harmonised or European Standards should be specified where they are applicable. Where no accepted standard or independent certification is offered, care should be taken with the manufacturer's specification to ensure that equipment will be satisfactory and reliable for the service required (see "References", paragraph 2.62). IEC or ISO standards, by agreement, are harmonised and adopted by those EEC and EFTA countries represented on the European Committee for Electrotechnical Standardisation (CENELEC).

## Uninterruptible power supplies

**2.40**  The considerable increase in sophistication and computerisation of equipment for specialised treatment and monitoring of patients has led to a greater demand for continuous and reliable power supplies. An AC "no break supply" that is continuous and unaffected by external circumstances can be provided by an uninterruptible power supply (UPS).

**2.41** UPS systems may be of one of the following types:

a.  battery-supplied bank of electronically controlled solid state inverters;

b.  battery-supplied direct current motor driving an AC generator;

c.  flywheel and engine combination driving an AC motor-generator.

**2.42** The capacity of a UPS battery for a computer installation may sometimes be based on the relatively short period between mains failure and the provision of a satisfactory supply from a standby generator. In practice many large computers will only function for a brief period without air conditioning, that is a load which must be supplied from an emergency generator in the event of mains failure. If the generator fails to start, the UPS battery merely provides a supply for the orderly shut down of the computer.

**2.43** A UPS for computer equipment can sometimes be justified on the grounds that it also provides a "cleaner", more stable and transient-free supply.

## Combined heat and power systems

**2.44** The purpose of combined heat and power (CHP) systems is to obtain greater energy utilisation from the fuel used in an engine-driven generator set. The latter's overall electrical efficiency does not usually exceed 30%. However, a CHP system using a similar generator could achieve an overall electrical and thermal efficiency up to 90% by utilising the previously wasted heat.

**2.45** Heat is generally recoverable at two temperature levels from the engine by gas/water or liquid/water heat exchangers:

a.  from the exhuast gas and moisture at elevated temperatures between 350°C and 450°C;

b.  from the cooling water circulating through the engine water jacket, and from the lubricating oil in intimate contact with the metal revolving and sliding parts of the engine at temperatures between 70°C and 90°C.

**2.46** For greater benefit, the engine should run as continuously as possible at full load. It follows that for CHP to operate at maximum efficiency the engine must run at a high load factor with a simultaneous requirement for the heat output. To achieve this continuous running, additional generating plant may be required for essential emergency supply support during breakdown or maintenance.

**2.47** Special tariff arrangements must be agreed with the regional electricity company for the export of surplus electrical energy and for the provision by them of potential electrical power, known as a "spinning reserve".

## Definitions

**2.48** **Emergency supply** – any form of electrical supply which is intended to be available in the event of a failure in the normal supply.

**2.49** **Essential service electrical supply** – the supply from an engine-driven AC emergency generator which is arranged to come into operation in the event of a failure of the normal supply and provide sufficient electrical energy to ensure that all essential functions of the health care and personal social services premises are maintained in service.

**2.50** **No-break supply** – a circuit continuously energised whether or not the normal supply is available.

**2.51** **Static inverter** – a semiconductor-based device to provide an AC output from a DC input without the use of moving parts.

**2.52** **Uninterruptible power supply (UPS) equipment** – a mains isolating power source. A static inverter or rotary motor-generator set providing a no-break, interference free, 50Hz AC sine wave output which may or may not be phase-related to the incoming 50Hz AC supply.

**2.53** **Essential circuits** – circuits of the essential services electrical supply so arranged that they can be supplied separately from the remainder of the electrical installation.

**2.54** **Generator set** – an engine-driven synchronous AC generator with exciter and other essential components.

**2.55** **Emergency lighting** – BS5266 Part 1:

    a. a maintained lighting system is one in which all emergency lighting lamps are in operation at all material times;

    b. a non-maintained lighting system is one in which all the emergency lighting lamps are in operation only when the supply to the normal lighting fails;

    c. a slave luminaire is one that is supplied from a central emergency power source and does not have its own internal secondary supply;

    d. a sustained luminaire contains two lamp systems, one energised from the normal supply and the other from a central battery supply in an emergency;

    e. escape lighting is that part of the emergency lighting system which is provided to ensure that escape routes are illuminated at all material times;

    f. rest mode is defined in European Standards whereby the emergency lighting may be switched off either manually or from a central point when the normal supply is switched off.

**2.56** **Stoichiometric combustion** – the ideal condition in the combustion of a fuel. The minimum oxygen content in air required to completely oxidise a given quantity of fuel and to obtain the maximum release of heat.

**2.57** **Excess air** (%) – the quantity of air (oxygen) supplied in excess of the requirements for stoichiometric combustion.

**2.58** **Sealed battery** – a battery that is totally sealed, with no provision for electrolyte replacement.

**2.59** **Vented battery** (unsealed) – a battery that is provided with a vent plug to release gases and the means to replace products of electrolysis.

## Standards and references

**2.60** The degree of correspondence between an IEC or ISO standard and a BSI standard is shown by the following symbols:

    $\equiv$   a standard identical in every detail;

    $=$   a technically equivalent standard, but with different presentation;

    $\neq$   a related but not equivalent standard.

**2.61**  CENELEC requires represented states to make national standards technically equivalent, that is harmonised (HD) or technically identical, that is, European (EN), to the IEC or ISO standard. British Standard specifications will eventually be fully replaced by CENELEC standards. All CENELEC standards are supported by the European Court.

## Standards

**2.62**  British Standards Institution specifications:

BS89, 1977: 'Direct acting indicating electrical measuring instruments and their accessories'. ( ≡ IEC 51).

BS88, Part 1, 1982: 'Cartridge fuses for voltages up to and including 1000V AC and 1500V DC – General requirements'. ( ≡ IEC 269-1).

BS171, 1970: 'Power transformers'. (≠ IEC 76).

BS417, Part 2, 1987: 'Galvanised mild steel cisterns and covers, tanks and cylinders'. (metric units).

BS764, 1954 (1985): 'Automatic change-over contactors for emergency lighting systems'.

BS799, Part 5, 1987: 'Oil storage tanks'.

BS822, Part 6: 1964 (1988): 'Terminal markings for rotating electrical machinery'.

BS1710, 1984: 'Identification of pipelines and services'.

BS4417, 1969 (1981): 'Specification for semi-conductor rectifier equipments'. (≠ IEC 146).

BS2771, 1986: 'Electrical equipment of industrial machines'. (≠ IEC 204).

BS3535, 1962 (1987): 'Safety isolating transformers for industrial and domestic purposes'.

BS1361, 1971 (1986): 'Cartridge fuses for AC circuits in domestic and similar premises'. (≠ IEC 269).

BS1362, 1973 (1986): 'General purpose fuse links for domestic and similar purposes (plugs)'. (≠ IEC 269).

BS1363, 1988: '13A fused plugs and switched and unswitched socket-outlets and boxes'.

BS2869, 1988: 'Fuel oil for engines and burners for non-marine use'.

BS4343, 1968: 'Industrial plugs, socket-outlets and couplers for AC and DC supplies'. (≠ IEC 309).

BS1650, 1971: 'Capacitors for connection to power frequency systems'. (≠ IEC 70).

BS3951, 1969 (1977): 'Freight containers'. ( ≡ ISO 668).

BS4752, Part 1, 1977: 'Circuit-breakers'. (≠ IEC 157).

BS5304, 1988: 'Code of practice for safeguarding machinery'.

BS5000, Index: 'Rotating electrical machines of particular types or for particular applications'.

BS4999, Index: 'General requirements for rotating electrical machines'. (≠ IEC 34).

BS3938, 1973 (1982): 'Current transformers'. (≠ IEC 185).

BS3941, 1975 (1982): 'Voltage transformers'. (≠ IEC 186, 358).

BS4196, 1981 (1986): 'Sound power levels of noise sources'. ( ≡ ISO 3740/6).

BS5266, Part 1, 1988: 'Code of practice for the emergency lighting of premises'.

BS5378: 'Safety signs and colours'. (≠ ISO 3864).

BS5424, Part 1, 1977: 'Contactors up to and including 1,000V AC and 1,200V DC'. (≠ IEC 158-1).

BS5514, Parts 1/6, 1988: 'Reciprocating internal combustion engine performance, etc' . ( ≡ ISO 3046).

BS6231, 1981: 'PVC insulated cables for switchgear and control gear wiring'.

BS6346, 1989: 'PVC insulated cables for electricity supply up to and including 3300V between phases'.

BS6132, 1983: 'Code of practice for safe operation of alkaline cells'.

BS6133, 1985: 'Code of practice for safe operation of lead-acid cells'.

BS6260, 1982, 1988: 'Open nickel-cadmium prismatic rechargeable single cells'. (≠ IEC 623)

BS6290, 1982, 1988: 'Lead-acid stationary cells and batteries'.

BS5000, Part 3: 1980 (1985): 'Generators to be driven by reciprocating internal combustion engines'.

## References

'Notes of guidance for the protection of private generating sets up to 5MW for operation in parallel with the Electricity Board's Distribution Network', (ETR No. 113, 1989 – The Electricity Association)

'Recommendations for the connection of private generating plant to the Electricity Board's Distribution Systems', (G59, 1985 – Electricity Association)

'Limits for Harmonics in the UK Electricity Supply System', (G 5/3, 1976 – Electricity Association)

'Regulations for Electrical Installations (16th edition)', Institution of Electrical Engineers (IEE)

'Lighting guide for hospitals and health care buildings', Chartered Institution of Building Services Engineers

'Firecode' suite of documents (NHS Estates)

'Code of practice for reducing the exposure of employed persons to noise', Health and Safety Executive

HN (76) 126 – 'Noise control'

IM/17 – 'Code of practice for gas engines', British Gas

American National Standards Institute/Underwriters Laboratory (ANSI/UL) – 'Automatic Transfer Switches', 1008, 1983

IEC 947-6-1 – 'Low voltage switch gear: Automatic Transfer Switches'

ISO 3046, parts 1/6 – 'Reciprocating internal combustion engine performance'

ISO 8528: to replace ISO 3046 after harmonisation

National Health Service Model Engineering Specification C44 – 'Diesel Engine Driven Automatic Stand-by Generator Sets'

# Symbols for diagrams

**2.63**   The following symbols for diagrams are recommended by BS1192, Part 3, 1987:

| | |
|---|---|
| ⅄ | Socket-outlet |
| ⊗ | Indicator lamp |
| | Switch or isolator |
| | Isolator used as bus-section switch |
| | Manually operated changeover switch |
| | Fuse link |
| | Fuse |
| | Fuse switch |
| | Circuit breaker |
| | Contactor |
| | Electric motor starter |
| | Changeover contactor comprising two contactors with mechanical and electrical interlocks |
| | Relay or contactor operating coil |
| | Push-button switch |
| | Self-contained escape/standby luminaire |
| | Double wound transformer |
| | Storage battery |
| | Horn or hooter |
| Ⓖ | Synchronous AC generator |
| | Control panel |
| avr | Automatic voltage regulator |
| | Connection to earth conductor or earth electrode |

# List of figures

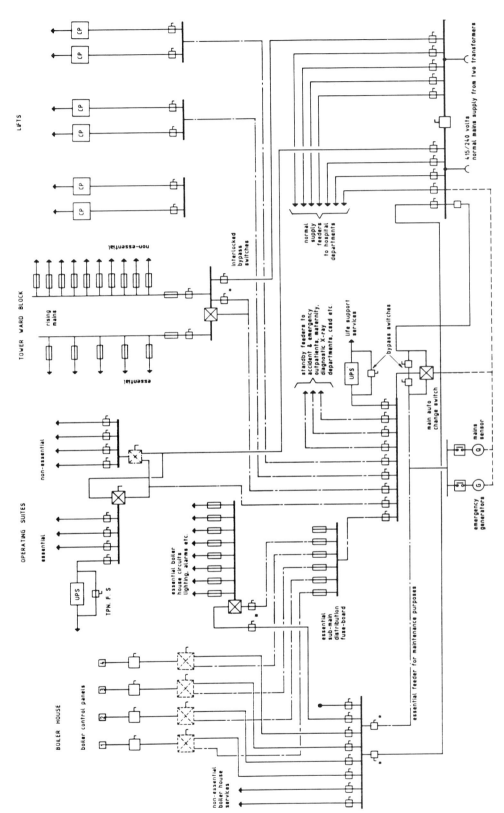

Figure 1. Typical distribution of standby and normal supplies for a district general hospital

\* Switches interlocked to prevent interconnection of the two supplies

NOTES

(1) Where economically practicable interconnecting cables should be provided between the sub-distribution centres of hospital departments to provide alternative feeders. Interlocked isolators should be provided to prevent interconnection of the normal and standby supplies.

(2) Single-line representation is used in this diagram for single and 3-phase distribution.

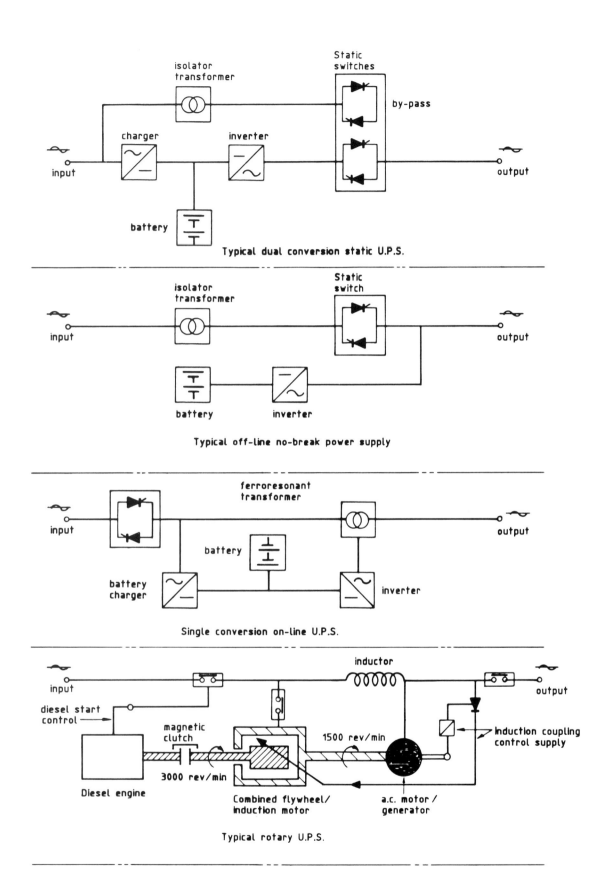

**Figure 2.** Typical U.P.S. schemes for emergency systems.

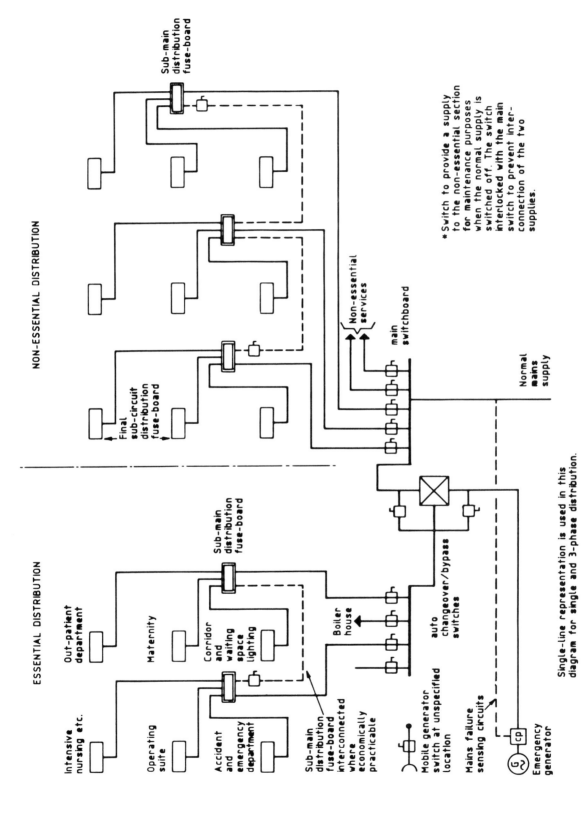

Figure 3. Typical distribution with an auto change-over contactor and by-pass switch for a small hospital.

Single-line representation is used in this diagram for single and 3-phase distribution.

*Switch to provide a supply to the non-essential section for maintenance purposes when the normal supply is switched off. The switch interlocked with the main switch to prevent inter-connection of the two supplies.

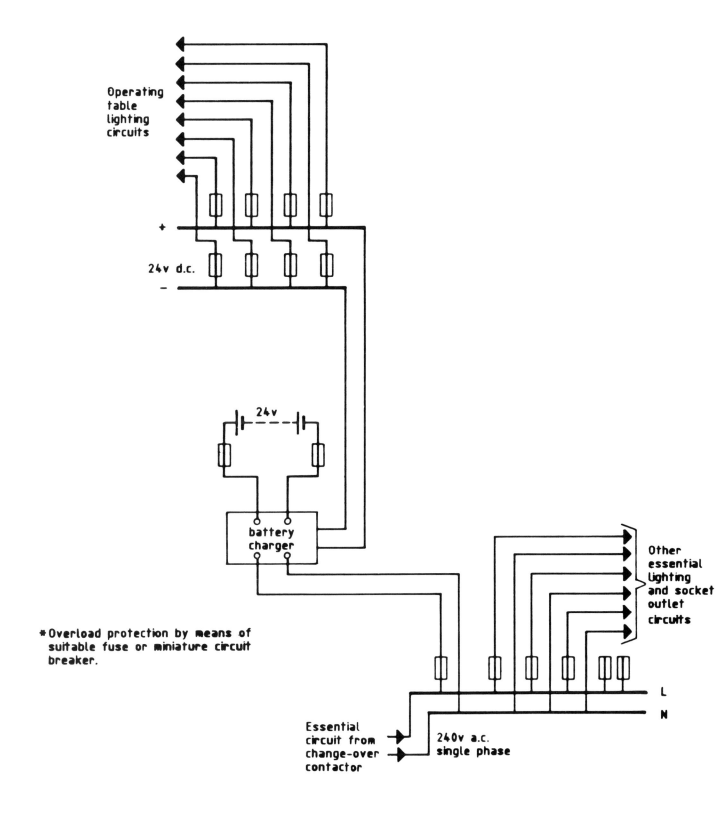

**Figure 4. Essential lighting and socket outlet supplies for operating departments.**

MOTOR STARTING CURRENT (PER UNIT) / GENERATOR FULL LOAD CURRENT (PER UNIT)

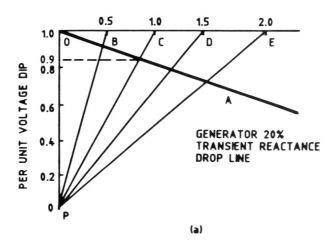

(a)

| TYPE OF MOTOR | METHOD OF STARTING | K VALUE |
|---|---|---|
| S.C. | D O L | 7 |
| S.C. | STAR / DELTA | 2.5 |
| S.C. | AUTO. T. | 3 - 4 |
| S.R. | ROTOR RESIST. | 1.5 |

MOTOR STARTING KVA = K x $\sqrt{3}$ V.I$_m$

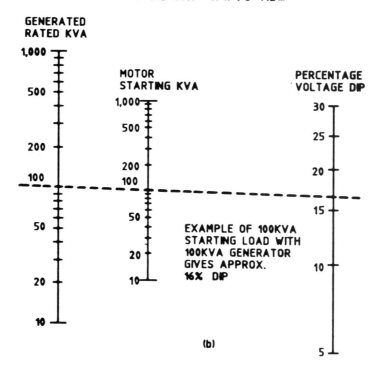

(b)

**Figure 5**: (a) Construction to find value of voltage dip

(b) Nomogram giving voltage dips for various motor starting loads

43+/SOL/TG/1

| Battery Type | Characteristics under abnormal Charging conditions | | | Maintenance | | | Expected of Life in years | Cost Ratio per Unit | Remarks | Applications |
|---|---|---|---|---|---|---|---|---|---|---|
| | Prolonged Overcharge | Prolonged Discharge | No normal discharge | Topping up | Change State | Cleanliness | | | | |
| Lead Acid Sealed | Capacity reduced because of plate shedding. Life reduced | Negative plate hardens. Life reduced. | Reduces life | Not required | Not applicable | Optional | 3 to 5 | 0.74 | | Fire alarm systems up to 40 bells. Self contained emergency lighting luminaires |
| Open vented High performance Pasted Flat Plate | Ditto | Negative plate hardens. | Reduces life | Required every 4 or 12 weeks depending on type | Check weekly with hydrometer | Necessary | 5 to 8 or 10 to 12 depending on type | 1.00 | | Standby Generator starting or Emergency lighting and Fire Alarm Systems depending on type |
| Open vented High performance Tublar | Ditto | Negative plate hardens. | Reduces life | Required every 8 weeks | Check weekly with hydrometer | Necessary | 4 to 6 | 0.86 | | Electric tugs and trucks |
| Plante | Life reduced to lesser extent | Ditto | Reduces life | Required every 12 to 24 weeks | Check weekly with hydrometer | Necessary | 15 to 20 | 1.35 | Capacity maintained until plate collapses | Emergency lighting. Fire Alarm Systems and Standby Generator Starting |
| Nickel Cadmium Sealed | Reduced capacity and life | Reduced capacity and life | Reduces life but less than Lead Acid | Not required | Not applicable | Optional | 3 to 7 | 1.80 | | Fire Alarm systems up to 40 bells. Self contained emergency lighting luminaires |
| Open Vented Heavy Duty | Satisfactory providing plates immersed | Satisfactory providing subsequent recharge | Satisfactory up to rear end of life | Should be checked every 26 weeks | Check monthly with voltmeter preferably on discharge | Preferable | 20 to 30 | 2.20 | | Standby Generator starting Switchgear operation |
| Open Vented Light Duty | Satisfactory providing plates immersed | Satisfactory providing subsequent recharge | Satisfactory up to rear end of life | Should be checked every 26 weeks | Check monthly with voltmeter preferably on discharge | Ditto | 20 to 30 | 1.90 | | Emergency lighting and Fire Alarm Systems |

Expected of Life in years (note for last row): These figures assume maintenance has been correctly carried out and battery has been charged and discharged normally or as recommended

Cost Ratio per Unit (note for last row): Cost ratio per unit is related to Lead Acid open vented. Example: Lead acid open vented Plante approx 35% more expensive

Fig 6 - Summary Table for Battery selection with Constant Voltage Charger

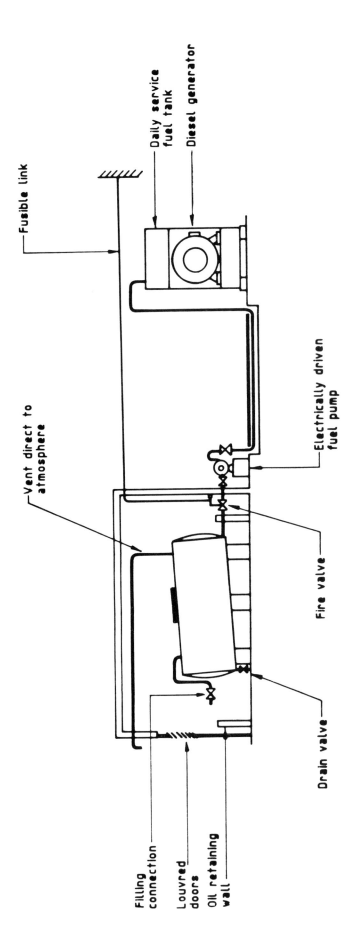

**Figure 7.** Typical bulk diesel oil storage installation.

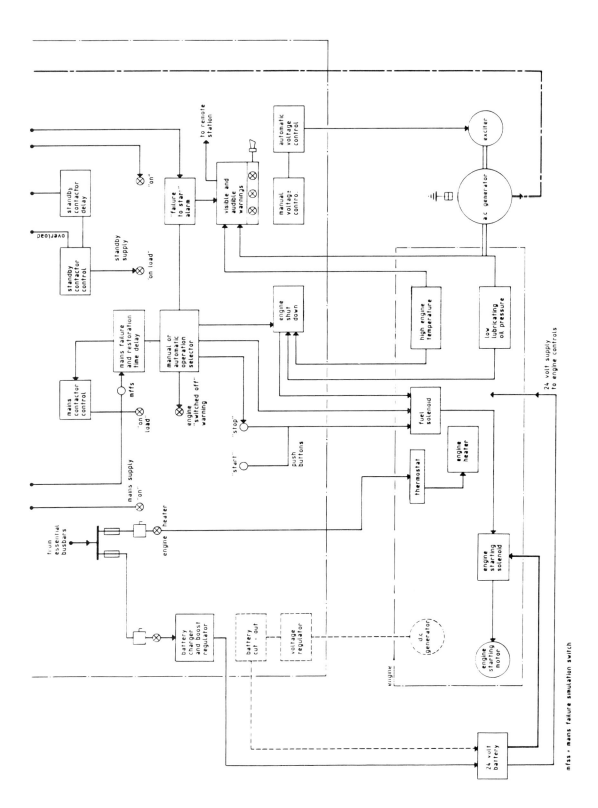

Figure 8. Typical diesel generator controls for automatic mains failure operation

mfss = mains failure simulation switch

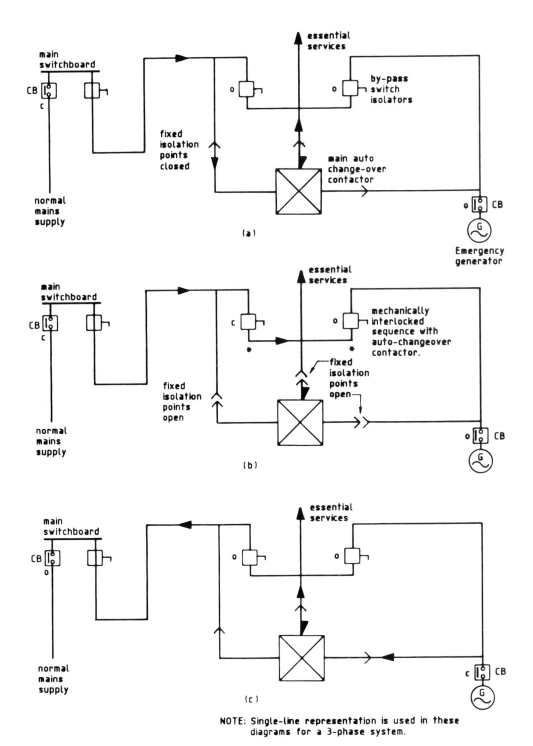

**Figure 9.** Maintenance switching arrangements for the emergency generator auto change-over switch.

(a) Normal operation.
(b) Change-over switch by-passed for maintenance by isolation. Supply normal.
(c) Standby generator used to supply the essential load.

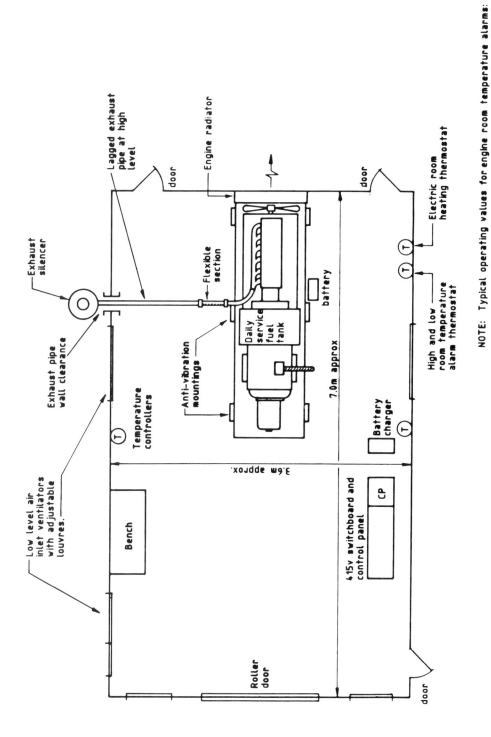

NOTE: Typical operating values for engine room temperature alarms:

high temperature    35 C
low temperature      7 C

Remote indicator lamps and audible alarm positioned at a permanently manned station, for example in the boiler house.

**Figure 10.** Typical diesel generator installation (100kW).

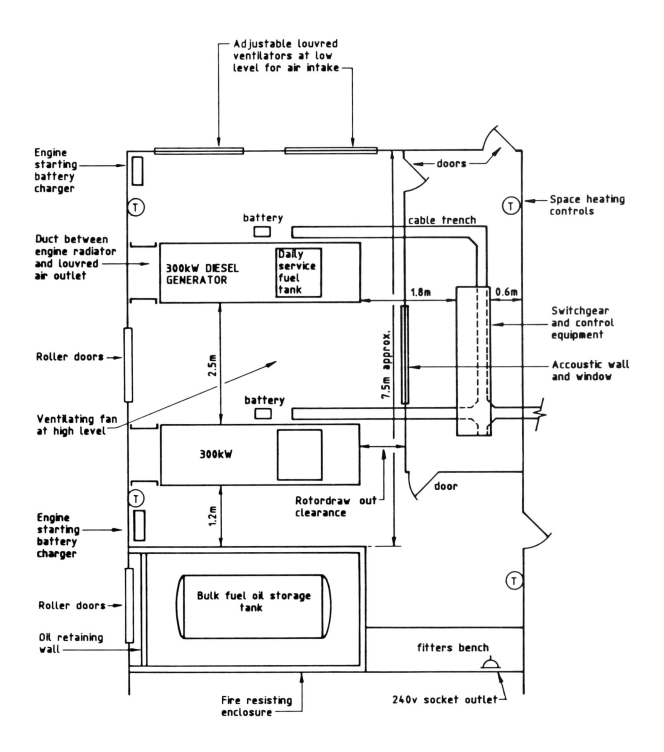

Figure 11. Typical layout for two 300kW diesel generator sets.

NOTE: Layout not to scale.

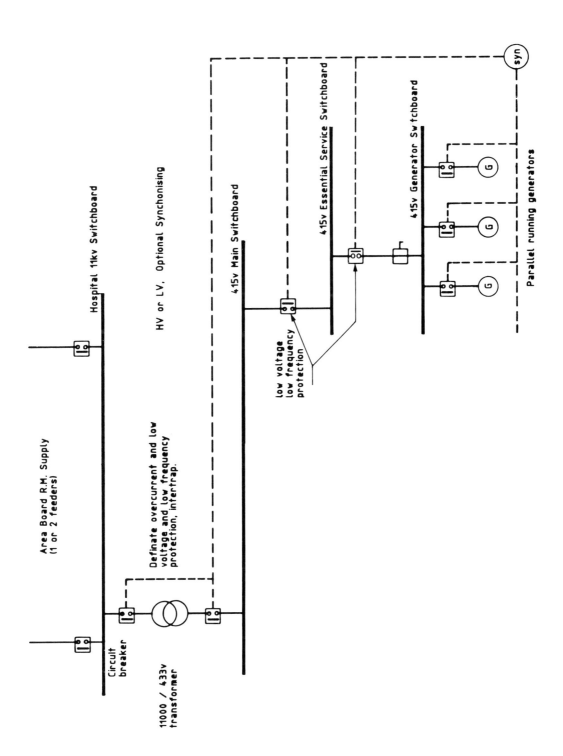

**Figure 12.** Typical arrangement for hospital main switchboard running in parallel with the Area Board Supply.

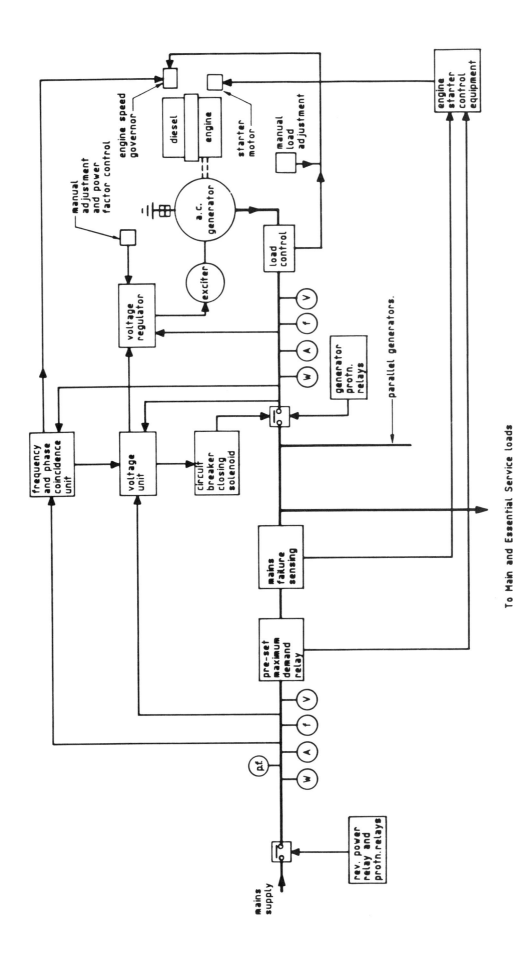

**Figure 13.** Controls for automatically synchronising a diesel generator running in parallel with the mains supply.

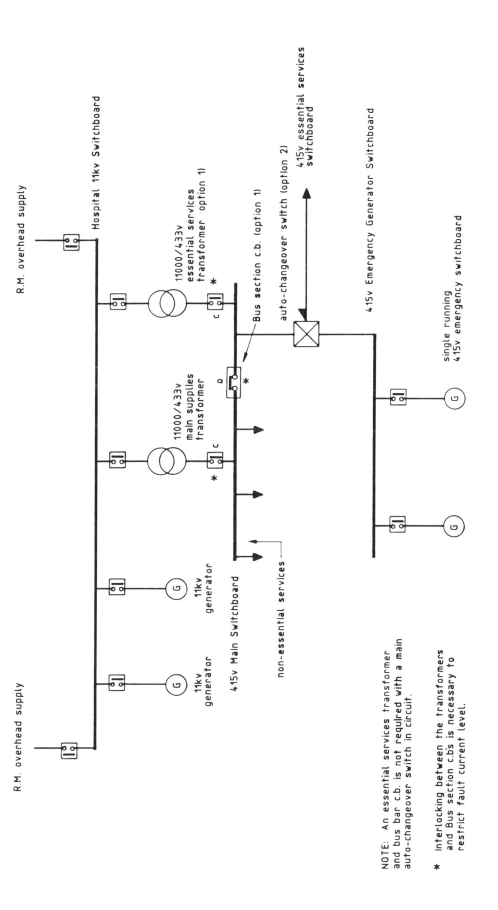

**Figure 14.** Arrangement for a large hospital where some of the electrical power is generated at 11000 volts.

R.M. overhead supply

Hospital 11kv Switchboard

R.M. overhead supply

11000/433v essential services transformer (option 1)

11000/433v main supplies transformer

11kv generator

11kv generator

415v Main Switchboard

non-essential services

Bus section c.b. (option 1)

auto-changeover switch (option 2)

415v essential services switchboard

415v Emergency Generator Switchboard

single running 415v emergency switchboard

NOTE: An essential services transformer and bus bar c.b. is not required with a main auto-changeover switch in circuit.

* Interlocking between the transformers and Bus section c.bs is necessary to restrict fault current level.

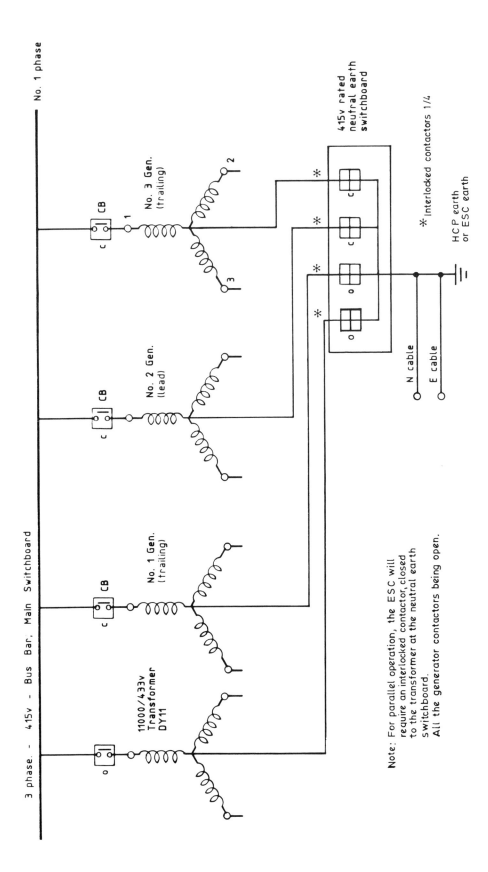

**Figure 15.** Typical arrangement showing No. 2 generator solid earth connection (TN-S) at the neutral earth switchboard for island operation.

# 3.0 Lighting and power supplies

## Emergency services – lighting requirements

### General

**3.1**   There are two main categories of emergency lighting, "escape" and "standby" as defined in BS5266, Part 1, 1988 'Emergency lighting'. Standby lighting is further subdivided into two grades, (a) and (b) – see also paragraph 3.11. Refer to Schedule, item 4 and Chartered Institution of Building Services Engineers (CIBSE) – 'Lighting Guide for Hospitals and Health Care Buildings' and/or 'Firecode' for Nucleus hospital requirements as applicable.

**3.2**   In the Nucleus design the emergency lighting requirements vary considerably from those of older health care and personal social services premises. This is a result of the Nucleus building design incorporating discrete one hour fire-resisting standard containment areas, more natural light into the building and the layout of main street escape routes. Agreement and guidance of the local fire authority should be obtained during the planning and installation phases of construction.

### Escape lighting

**3.3**   The escape luminaires can be of either the maintained or non-maintained variety. They can be powered by a suitable battery supply connected by auto-changeover switch or they can be adequate self-contained luminaires. In less hazardous locations standby luminaires are permitted when supplied within 15 seconds by the emergency generator.

**3.4**   Escape lighting is required to define easy to follow exit paths, especially in premises of old, labyrinth-type design and those with satellite buildings with complicated exit paths. Requirements for lighting are as follows:

   a.   at each exit door;

   b.   within 2m of each change of direction and each intersection of corridors;

   c.   within 2m of each staircase, giving direct light on each flight;

   d.   within 2m of any other change of floor level, for example a dais;

   e.   outside, and close to, each final exit;

   f.   within 2m of all fire alarm call points and extinguishers;

   g.   to illuminate all exit and safety signs;

   h.   floor level lighting in smoke stagnant spaces;

   j.   fluorescent floor markers.

**3.5**   Lighting should give a positive indication of escape routes to the outside of the premises and all illuminated signs giving such direction should be adequately lit. Light indicating an escape route should originate from more than one emergency luminaire.

**3.6**   After determination of the premises' exit locations, the positions of intermediate luminaires will depend upon the floor to ceiling available height, obstructions, the spacing factor of luminaires and the need for a minimum illumination of 0.2 lux at the centre line floor level. Fifty per cent of the floor area for all escape routes, up to a width of 2m, should be illuminated at 0.1 lux minimum.

**3.7** In general the higher the intensity of normal illumination, the higher should be the level of escape route illumination.

**3.8** Escape signs should conform to BS5378, Part 1: 1980 – 'Safety signs and colours', BS5499, Part 1: 1990 – 'Fire Safety Signs' and Safety Signs Regulations 1980, Statutory Instrument 1980/1471.

*Safety Signs Regulations (Northern Ireland), SR 1981 No 352*

**3.9** Luminaires and wiring used in defined escape routes must comply with the non-flammability requirements of BS4533: Section 102.22, 1981 – 'Luminaires for emergency lighting', and BS5266, Part 4, 1990 – 'Centrally supplied emergency lighting systems' and BS6387: 1983 – 'Fire resistant cables'.

## Standby lighting

**3.10** Standby lighting energised from the essential services supply is provided to enable normal activities to continue during loss of normal lighting supply. In critical work areas such as operating theatres, delivery rooms and high dependency units, the quality and intensity of standby lighting should be equal or nearly equal in illuminance to that of the normal lighting at the task points. General standby lighting in the surrounding areas may be reduced if considered acceptable.

**3.11** Two grades of standby lighting are recommended, as defined in the CIBSE Lighting Guide – 'Hospitals and Health Care Buildings' and BS5266, Part 1 – 'Emergency Lighting':

a. **Grade (a)** – lighting of intensity and quality equal or nearly equal to that provided by the normal lighting;

b. **Grade (b)** – a reduced standard of lighting, about one third to one half of normal lighting intensity, sufficient to enable general health care and personal social services premises activities to continue.

## Typical standby lighting

**3.12** The schedule opposite is intended as a general guide to typical grades of standby lighting requirements. Reference should be made to the Chartered Institution of Building Service Engineers' 'Lighting Guide for Hospitals and Health Care Buildings' for more specific design details.

*Schedule 1.* Typical grades of standby lighting in departments/locations

| Department or location | Area | Grade of lighting |
|---|---|---|
| Operating department | } critical working areas<br>general working areas | grade (a)<br>grade (b) |
| Maternity department | } critical working areas<br>general working areas | grade (a)<br>grade (b) |
| Accident and emergency department | } critical working areas<br>general working areas | grade (a)<br>grade (b) |
| Administrative department | general working areas | grade (b) |
| Out-patient department and special treatment clinic | } critical working areas<br>general working areas | grade (a)<br>grade (b) |
| Dispensary | dispensing areas | grade (a) |
| Patient care areas | general working areas | grade (a) |
| Bed areas in wards, etc | general working areas | grade (b) |
| Treatment rooms | general working areas | grade (a) |
| Drug stores and DDA cupboards | general working areas | grade (a) |
| Kitchen | general working areas | grade (b) |
| Psychiatric department Treatment areas | } critical working areas<br>general working areas | grade (a)<br>grade (b) |
| Pathology department | } essential working areas<br>general working areas | grade (a)<br>grade (b) |
| Rehabilitation department | treatment and general areas | grade (b) |
| Diagnostic X-ray department | critical working areas | grade (a) |
| Radiotherapy department | } critical working areas<br>general working areas | grade (a)<br>grade (b) |
| Mortuary and postmortem room | general working areas | grade (b) |
| Kitchens and dining rooms | general working areas | grade (b) |
| Lifts, landings and lobbies | | grade (b) |
| Laundries | general working areas | grade (b) |
| Sterile supplies | general working areas | grade (b) |
| Boiler house and plant rooms | working areas | grade (b) |
| Generator room | | grade (a) |

# Arrangements of emergency lighting circuits and luminaires

### Standby lighting circuits

**3.13**   Where maximum standby lighting (grade (a)) is required under emergency conditions, all circuits should be connected to the essential services supply.

**3.14**   In areas where a reduced intensity of standby illumination is acceptable (grade (b)), selected circuits only should be connected to the essential supply.

**3.15**   Where continuous or task lighting is required in dark, unsafe or emergency working areas, maintained emergency luminaires should be provided. These should be of adequate luminosity to bridge extended blackouts or losses in illumination due to delays in the stationary emergency generator starting, for example inside the diesel generator room, switchroom, kitchens, plant rooms or operating theatres. Non-maintained emergency luminaires may be used in areas falling outside these safety or emergency working categories.

**3.16**   Where economically viable the essential and normal mixed small power and lighting in each area should be energised by cable having segregated supply routes to safeguard against total loss of lighting supply.

### Escape lighting circuits

**3.17**   Definitions are given in paragraph 2.55.

**3.18**   All escape luminaires, except self-contained luminaires, should be energised by segregated and dedicated fire resistant emergency circuits. The wiring circuits should be routed to avoid, where possible, passing through areas of high fire risk or high fire loading. These escape luminaires should be controlled by an automatic changeover switch, jointly supplied as slave maintained lighting, or supplied only as slave non-maintained lighting from either the essential services supply or a battery source. The response time should be to operate immediately from a battery or within 15 seconds from an emergency generator, subject to the agreement of the local fire authority. Item 3.3 refers.

### Battery operated escape lighting

**3.19**   A storage battery should be of suitable rating and capacity to supply the total load energy to the circuits supplying emergency escape lighting for a period of three hours, as required in BS5266, Part 1, 1988 – 'Emergency Lighting'. However, a period of one hour applies if emergency power generated lighting is also provided in areas occupied by only ambulant patients, staff or public within the maximum 15 second delay requirement.

**3.20**   Provision of a 24V low capacity battery of at least one hour duration is required as an emergency alternative supply for main and satellite operating theatre table luminaires used in all forms of acute surgery. Figure 4 (on page 19) shows a typical circuit. Fully charged battery powered hand torches should always be available.

**3.21**   The general lighting in an operating department should be connected to the essential services supply.

**3.22**   BS5266, Part 1, 1988 gives guidance on the choice of capacity for battery power packs.

**3.23**    In health care and personal social services premises where the AC emergency generator sets are also used for combined heat and power functions, connection of escape lighting to a segregated three-hour battery supply by operation of an auto-changeover switch is recommended.

## Luminaires

**3.24**    Self-contained emergency luminaires fed from the normal services lighting supply may be used in all areas.

**3.25**    Self-contained emergency luminaires are available in maintained and non-maintained versions as escape lighting and escape lighting exit signs with battery packs of one hour or three hours' duration.

**3.26**    Voltages to emergency luminaires vary from 24V, 50V, 110V and 240V with output ranges of 4 watts, 8 watts and 16 watts for fluorescent lamps and 8 watts, 15 watts and up to 100 watts for incandescent tungsten lamps.

**3.27**    High frequency fluorescent self-contained surface luminaires have been introduced for maintained use. These have one- or three-hour batteries with inverters to provide the HF/AC supply on loss of normal supply. They operate at one-third to one-half illumination level when in the battery mode, and may be connected to the normal supply as a switchable supply in working areas acceptable for energy economies.

# Emergency services – Power equipment requirements

## Socket-outlets

**3.28**    As a general guide, in any area where standby lighting is provided to enable normal activities to be carried out during loss of normal supply, all power socket-outlets conforming to BS1362 or IEC harmonised standards should be connected to the essential/unified circuit supply. This arrangement is recommended to simplify and standardise the electrical installation and choice of socket-outlet positions on the walls. Only essential equipment required at the time should be plugged into socket-outlets connected to essential supply circuits or remain in service during periods when only the emergency AC generators are providing the power supply to the premises.

**3.29**    Essential supply socket-outlets are recommended to be labelled differently from those connected to segregated or unified non-essential supplies.

## Special sockets or locations

**3.30**    In locations where an assembly of related low powered life-support equipments are in regular use, dedicated essential supply or no-break supply ring or radial main multi-socket arrangements should be provided for each assembly. The installation must conform to the IEE document 'Regulations for Electrical Installations' as a TN-S system or, in special intercardiac or wet areas, as an IT system. (IEC 364-7-710 refers.)

**3.31**    Where standby lighting is provided for the well-being and/or security of patients, such as in long stay, geriatric or psychiatric health care and personal social services premises, socket-outlets should only be connected to the essential supply where the socket-outlets are required to supply essential equipment.

## Sterilizing equipment

**3.32**  Electrical services, including alarms, automatic controls and ancillary circuits essential for the safe operation of sterilizing equipment should be connected to the essential supply.

## Dirty utility rooms

**3.33**  An essential power supply should be provided for bedpan washers and macerating machines.

## Patient care and life-support equipment

**3.34**  All special patient care and patient monitoring equipment should be supplied from the essential supply and given first priority in the event of loss of normal supply. In specifically designated areas where life-support power equipment is in use the need for uninterruptible power supplies (UPS) with at least a one hour duration battery and electronic and/or manual bypass switch should be considered. The UPS input should be connected to the essential supply circuit. (See paragraphs 3.68-3.89.)

## Blood banks etc

**3.35**  Blood banks and other clinical refrigeration equipment are sufficiently thermally insulated to retain low preserving temperatures for several hours. As a precaution it is recommended to supply this type of equipment from the essential supplies. Where haematology analysis is conducted it may be necessary to provide a small UPS with battery support to prevent damage to tests in progress during the change over from normal to emergency AC generator supply.

## Deep-freeze refrigerators and food stores

**3.36**  These refrigerators will normally operate within a range of minus 12°C to minus 23°C and be fitted with a temperature alarm to give warning when the refrigerator temperature approaches the upper safety limit. As the electrical loading is comparatively small they may be supplied from essential circuits as a safeguard against prolonged supply interruptions.

## Ward kitchens

**3.37**  Refrigerators in food preparation rooms, special care baby units and children's wards should be supplied from essential circuits. Heat regeneration trolleys for cook-chill foods should have the essential power demand controlled to prevent sudden overloading of the emergency generators.

## Cold stores

**3.38**  These usually operate within a temperature range of 0°C to minus 2°C and are normally provided with sufficient insulation to prevent undue rise in temperature during power failure. It may, however, be desirable to have an alternative supply from the emergency plant available where this can be conveniently arranged.

## Post mortem room

**3.39**  Refrigerating plant for body chambers should be supplied from an essential circuit.

### Kitchens

**3.40**   The use of essential electricity supplies will not normally be justified for electric cooking. Where the main kitchen equipment is electrically heated it may be considered advisable to provide alternative non-electric heating facilities as a safeguard.

### Boiler plant

**3.41**   Essential supplies should be restricted to those items of auxiliary plant that are necessary to enable the plant with its associated alarms, safety controls and ancillary circuits to continue to provide its primary heating function in the event of loss of normal supply. The essential auxiliaries include the following:

a.  mechanical coal stokers or oil burner pressure pumps;

b.  oil storage tank and fuel pipework trace heating;

c.  forced and induced draught fans and other mechanical ventilation equipment necessary for safe boiler operation;

d.  boiler feed pump and condensate pump;

e.  heating and flue gas control circuits and/or air compressors;

f.  circulating water pumps for space heating and hot water services;

g.  heat exchanger stations.

### Diagnostic X-ray machines

**3.42**   Where X-ray apparatus is used in diagnostic rooms at least one room must be supplied from the essential circuits. Socket-outlets on essential circuits should be provided to permit long duration clinical procedures on a patient connected to a life-support machine to continue. The X-ray machine may be required to monitor clinical progress, but should not be used for high power film work.

### Ventilation

**3.43**   Essential supply facilities should be provided for ventilation plant where mechanical ventilation is required for clinical reasons.

### Central piped medical gases

**3.44**   This section includes medical compressed air and medical vacuum installations (see also Health Technical Memorandum 2022 – 'Piped Medical Gases' (in preparation)). The associated electrical services of piped medical gases, including any safety controls and ancillary circuits, should be connected to essential circuit supplies. Alarms should have a no-break battery supply. It is desirable that the motor compressor installation should have an auto-start selection mode of pressure control for motor contactor operation to ensure re-starting of motors after operation of the changeover switch. The provision of a time-delayed auto-start mode of control will require a local/remote/auto control selector switch mounted on the starter panel. The stop push button located at the motor should be of the shrouded type to prevent inadvertent operation.

### Water supplies

**3.45**   Where electric motor driven pumps are used to maintain essential water supplies it is necessary to provide suitable time delayed auto-start control arrangements for the pump electric motors to be reconnected to the essential circuit supplies.

**3.46**  Essential services equipment should not be dependent on cooling water directly from the mains water supply.

## Essential supporting services

**3.47**  Central supporting services, such as laundries and food preparation cook-chill facilities that operate on a 24-hour a day basis, should be able to maintain an essential service. These services should be provided with 100% emergency AC generator set capacity to replace total loss of normal power supply for an extended period.

## Sewage disposal

**3.48**  Where sewage does not directly flow into the municipal sewerage system, but is processed partially or wholly at a health care and personal social services premises sewage farm, the installation should be protected with an alternative essential services supply.

## Lifts

**3.49**  Where lifts are provided for the movement of patients, it is recommended that at least one lift of orthopaedic bed capacity is installed in each section of the hospital. All such lifts should be connected directly to the essential services supply.

**3.50**  The ability of the generator, essential services system and electronic control circuitry to withstand harmonic or transient voltage effects injected by the lift motor should be established.

**3.51**  Emergency self-contained luminaires, of minimum 1 watt rating, should be provided in all lift cars. The illumination should be adequate to alleviate distress in passengers.

## Telephone exchanges, communications, security equipment and fire alarms

**3.52**  The following battery chargers and equipment should be supplied by a switchboard or distribution board energised directly from the essential services supply:

a.  telephone exchanges;

b.  communication systems used for direction and control during external and hospital emergencies;

c.  security equipment for safety of personnel and protection of property;

d.  fire alarm equipment.

## Nurse call and staff call equipment

**3.53**  The electrical load for this equipment is very small and there is no reason why it may not be connected directly to the essential services supply.

## Computer systems and data communications equipment

**3.54**  Computer supply systems should be no-break and should ensure an interference-free input voltage.

**3.55**  Ventilation equipment supplied from the essential services supply should be provided for all computer equipment suites. A high degree of filtration of the admitted air together with close limits on humidity and temperature control must be provided.

**3.56**   Inlet and outlet ventilation ductwork to computer equipment suites must be protected by auto-close fire dampers controlled by smoke- and heat-sensitive detectors.

**3.57**   Electrical services supplying computer suite ventilation equipment and luminaires should be provided with a means of shutdown isolation at the main point of exit from the computer suite.

**3.58**   An alarm panel should be provided to monitor UPS and ventilation control excursions.

## Segregation of essential and non-essential circuits

### Fire and segregation

**3.59**   Where the same cable route is used for essential and non-essential circuits in an installation, all possible efforts should be taken to provide cable segregation and to reduce the possibility of a single cable fault damaging both circuit cables.

**3.60**   In essential circuits, flexible fire-resistant cables tested to BS6387 may be used where essential and non-essential cable circuits cannot be segregated in ductwork or on traywork.

**3.61**   In important areas where there are a large number of essential sub-circuits, separate traywork should be used for the routing of essential and non-essential circuit cables.

**3.62**   Mineral insulated metal sheathed (MIMS) cable or any fire-resistant cable conforming to the category B test requirements of BS6387 – 'Performance Requirements of Cables Required to Maintain Circuit Integrity Under Fire Conditions' should be used for all escape lighting and fire alarm control circuits required to function under fire conditions. (See the 'Firecode' suite of documents.)

**3.63**   Complete segregation of category 3 escape lighting cable and fire protection wiring routes is required. IEE Regulations refer.

**3.64**   MIMS cable should not be routed on equipment subject to continuous vibration such as on engines or transformers, as vibration may cause cracking of the metal sheath. Where MIMS is installed, the sheath materials are usually copper, but stainless steel may be used if preferred.

### Marking of essential circuits

**3.65**   Switchboards and distribution boards that marshal essential circuits should be clearly and indelibly marked to indicate their function. Switching devices and circuit equipment connected by the same cable should be identically labelled for function and plant item number.

**3.66**   All cables should be numbered and labelled at both the far and near ends adjacent to the terminating glands with the cable number corresponding to the entry in the cable schedule.

**3.67**   It is recommended that BS1710, 1984 – 'Identification of Pipelines' is used as the guiding colour code for identification of plant pipework and cables. The recommended colour for electrical services is light orange. It is also recommended that a "red on light orange" be used to indicate essential circuits and installations. For example, a red spot on a light orange background for switchboards and distribution boards, and a red band on a light orange band for conduit and trunking that contain only essential supply circuits.

## Uninterruptible power supplies

### General

**3.68**   Uninterruptible power supplies (UPS) may be required to provide a no-break single or three-phase supply to specific equipment during the period of blackout, from the loss of normal supply to the start-up and subsequent electrical power generation of the emergency generator. Without UPS even very short power breaks may cause loss or corruption of data in computer systems, laboratory tests may be jeopardised and the well-being of patients undergoing intensive care or surgical treatment may be threatened. A "UPS" to supply computer equipment can sometimes be justified on the grounds that it will also provide a stable and transient-free input voltage.

**3.69**   All specialised types of electronic equipment without or with less than substantial internal battery support and requiring a no-break supply should be supplied by a UPS. The UPS should be either from a permanent central source or from a portable desk top or side unit. The UPS units should preferably be rated for maximum power factor and provide output power as specified by the electrical/electronic equipment manufacturer. They should also ensure that non-linear load peak currents can be suitably absorbed without equipment stress.

**3.70**   A support battery of at least one hour full load capacity with an electronic and/or manual bypass switch should be provided. The choice of bypass switch depends on the level of service required by medical or computer staff. The UPS should be connected to the essential services supply.

**3.71**   The capacity of a UPS battery for a computer installation may sometimes be based on the relatively short period between mains failure and the provision of a satisfactory supply from a standby generator. In practice many large computers will only function for a brief period without air conditioning, that is, a load which must be supplied from an emergency generator in the event of mains failure. If the generator fails to start, the UPS battery merely provides a supply for the orderly shut down of the computer.

### Combined heat and power and uninterruptible power supply

**3.72**   In some health care and personal social services premises electrical power is generated by a combined heat and power (CHP) system provided by retro-fitted emergency generators. Where this represents a large part of the total energy requirement, a UPS should be utilised to provide a no-break supply for grade (a) standby lighting and life support equipment in the event of CHP breakdown.

**3.73**   UPS should be considered as a source of supply to overlap any short or unexpected long periods of emergency generation (see Figure 1 on page 16).

### Types of uninterrupted power supply

**3.74**   UPS comprises two basic types: battery connected static inverters and rotary motor-generator assemblies. Neither are cheap installations. Overall efficiencies vary from 70% to 90%. The rotary type is the more expensive, but becomes competitive with the static type in large central data processing systems above 200kVA rating.

**3.75**  A typical static inverter UPS (see Figure 2 on page 17) may consist of:

a.  single conversion – a constant voltage, line filtering ferroresonant transformer supplied through a solid state switch and connected directly to the load. The transformer is in parallel with a battery charger, floating battery and inverter. On mains power failure the solid-state switch opens. This permits the battery to supply the load via the DC/AC inverter through the ferroresonant transformer;

b.  double conversion on-line – an input rectifier battery charger connecting to both a battery of suitable capacity on float charge and a DC/AC output inverter. A no-break supply.

To permit maintenance, or in the event of a complete UPS failure, the whole UPS should be bypassed by an isolating transformer in series with a solid-state bypass switch or a mechanically operated overall bypass switch fed from another point of essential services supply.

A further safeguard would be to have a second 100% rated UPS as passive standby or sharing load in parallel, with an auto isolating switch, in the event of failure of the duty UPS.

**3.76**  A typical rotary UPS consists of:

a.  an input rectifier with a floating battery supplying a DC motor/AC generator set. The DC motor speed is regulated by a time clock controller, comparing the supply and generator frequencies to give a constant output frequency and a no-break supply;

b.  a universal AC machine, operating in parallel with the normal supply, drives a combination flywheel/induction motor which augments the flywheel speed to almost twice synchronous speed. On loss of normal supply, the high speed flywheel magnetically slips, maintaining a constant speed torque to the universal AC machine now driven as a generator, to provide the initial supply. A diesel engine concurrently runs up to speed and magnetically couples to the flywheel to drive the universal AC machine at constant speed as an emergency generator to replace the normal supply.

### Noise

**3.77**  The location of highly rated UPS equipment should be carefully selected, as the audible noise level increases as the transistor/thyristor switching speed is reduced with increased rating. The noise distribution curve in decibels should be obtained from the manufacturer. Typical switching speeds of 15 to 20kHz give noise levels of the order of 52dBA for ratings of 3 to 20kVA. At 2kHz, a noise level of 60dBA for ratings up to 80VA should be expected. Reference should be made to HN(76)126 – 'Noise Control' and Appendices HDN4 (Noise Control).

### Waveform

**3.78**  Present designs of static double-conversion UPS use pulse width modulation (PWM) inverters. This type of inverter shapes an approximate sine wave from DC square wave impulses by variation of the impulse width. The impulses are generated at high frequency for minimal harmonic distortion, which contributes to the reduction of the capacity of inverter output filters. The overall size of the UPS is reduced with PWM as no large transformers are involved. Ferroresonant transformers are used in single conversion UPS. These give improved voltage regulation and line filtering with increased efficiency but are limited for ratings up to 240V, 20kVA.

**3.79**   In comparison with the above, the rotary UPS AC output generator produces an almost pure sine wave for the largest rating required.

**3.80**   Load-generated voltage distortion can be reduced by providing low-pass filter traps at the inverter output terminals. This filters harmonics from the voltage output waveform and helps to reduce any over-rating of the UPS required to match the load.

**3.81**   The load characteristics of electronic equipment such as computers tend to be non-linear and, as a result, it may be that the UPS required will be almost twice the kVa rating that would be expected from a linear load. The UPS manufacturer should analyse the equipment load waveform crest factor and harmonic distortion to determine the correct UPS rating.

**3.82**   Methods of earth connection at the UPS output side may differ to suit the load. This difference may be required to avoid mains-borne interference and transients, or, for example, for safety reasons in Medical Locations Group 2 where intercardiac surgery may be used. HTM 2007 (in preparation) refers.

### Harmonic distortion in uninterrupted power supplies and generators

**3.83**   Emergency generator output can suffer harmonic voltage distortion as a result of non-linear loads injected into the supply by the rectifiers in UPS and battery charger systems. Typical UPS values are third harmonic at 40%, fifth harmonic at 30% and seventh harmonic at 7%, all relative to the fundamental voltage. The effect of this harmonic voltage may be reduced by filters at the input terminals, provided by the inverter manufacturer. Guidance on the level of harmonic injection at the normal supply point of common coupling is given in Electricity Association Engineering Recommendation G5/3 (1976) – 'Limits of Harmonics in the UK Electricity Supply System'.

**3.84**   The generator winding reactance combined with each rectifier-generated harmonic current produces distortion of the generated voltage fundamental 50Hz waveform. It is recommended that total non-linear rectifier loads should not exceed 40% of the generator rating.

**3.85**   Non-linear voltages also affect the operation of AVRs, the indicated values of metering and the accuracy of speed measuring devices owing to the divergence from standard voltage form factors. It follows that, where installations have such non-linear loads, the following considerations should be made:

   a.  the rectifier/UPS may need a larger rating than apparent to supply the non-linear load and may need to be provided with frequency discrimination in the event of generator instability;

   b.  the generator rating may need to be increased owing to various rectifier-generated harmonic current heat losses within the generator windings. This is in addition to any extra rating required owing to motor starting currents;

   c.  the engine rating may need to be adjusted to suit the generator non-linear and dynamic start load current demanded, this being influenced by the engine load acceptance category.

### Protection of uninterrupted power supplies

**3.86**   The maximum energy that can be passed by a static PWM inverter UPS is 150% for one minute in the event of an overload, or a low impedance load fault, limited to a fuse clearance of 10 milliseconds without resorting to the AC static bypass switch. Where large capacitive filters are included in the inverter output circuit, the additional capacitance in these filters may be sufficient to assist the load short-circuit current for a period long enough to clear a suitably rated fast acting fuse or miniature circuit breaker. The high impedance electronic devices used in PWM designs with minimal output filters may not be of sufficient capacitance to operate the overcurrent protection devices in the faulty load circuit. Ferroresonant transformer UPS are normally current-limited to an overload capacity of 150% for 10 minutes.

**3.87** A three-phase output UPS should be able to cope with 100% unbalanced loads on each phase, while still providing a closely regulated output voltage and phase control.

**3.88** In high impedance UPS outputs, voltage or current-sensitive inverter protection should be considered. This can also be designed to synchronise the inverter output voltage with the supply voltage, then autoclosing the solid-state bypass switch into circuit with the load. The low impedance normal power supply will assist the load circuit protective devices. It is essential that suitable primary protection is installed on the supply side and graded to the load circuit.

### Installation

**3.89** The range of UPS available is extensive and spans desk top installations to large, single room assemblies. Where large installations are required, the best location is separate from, but near to, the data equipment and in the central area of a building. Owing to the concentration of equipment such as batteries and transformers, structural design must consider the extra floor loading, space for extra air-conditioning cooling and cableworks, and the lifting, access and possible removal of large equipment cubicles during and after construction.

# 4.0 Engines

## Emergency supply equipment

### General

**4.1**  The emergency supply required will be AC, with the same number of phases, rated voltage and frequency as the normal supply. In general, an engine-driven 415V, three-phase 50Hz AC generator set will be the most convenient and economical means of providing an emergency supply for the load demands of health care and personal social services premises.

**4.2**  The engine-driven generator set for the supply of essential circuit power and lighting is normally mounted on a permanent bed plate. It is designed for automatic starting in the event of a prolonged fall in normal supply voltage. A short delay of between 0.5 to 6 seconds is normally chosen at the voltage detector device to discriminate against a fall in normal voltage due to a voltage transient or auto reclose switching operation. When the chosen time delay confirms the loss of normal supply voltage the engine start is initiated. A time delay of up to 15 seconds is allowed in BS5266 between loss of normal supply and connection of the emergency generator to the essential services supply circuits. The AC emergency generator circuit-breaker should close when the generated voltage and frequency are at 95% of nominal values and before the auto-changeover load switch operates.

### Diesel engines

**4.3**  Generating sets of various rating are available designed for the generation of AC emergency electrical power. The choice of prime mover is usually the diesel engine, as it has a more rapid response time for load acceptance from a cold start (see paragraph 4.45).

**4.4**  The difference between makes of generator set can be arbitrary in performance and cost. Most set manufacturers can provide a specific make and design of engine to drive an AC generator at the request of the purchaser. Two basic types of diesel engine are available: truck and industrial. Owing to differing applications, the truck engine does not have the same proven long term life or reliability as the industrial engine, but it will normally be cheaper. Care must be taken when evaluating tenders.

### Gas engines

**4.5**  The main alternative choice of prime mover is the gas engine supplied from the gas mains. It is basically a modified and derated diesel engine with spark ignition and carburettor control. For a given rating it is bulkier than a diesel engine (see paragraph 4.31).

### Gas turbines

**4.6**  Gas turbines are available over a wide range of ratings for the generation of electrical power. However, even at the smallest rating of 500kW, health care and personal social services premises essential electrical loads are not likely to be sufficient to justify the cost of installation.

**4.7** Their main disadvantage is the time of one to two minutes required to reach full speed and the lack of ability then to accept large step loads up to full load when cold.

## Combined heat and power

*Electricity (Northern Ireland) Order 1991*

**4.8** Since the 1983 Energy Act there has been an upsurge in the use of combined heat and power (CHP) systems for efficient fuel utilisation and economy (see Appendix, page 78).

**4.9** With CHP a wider application of electrical generating plant in health care and personal social services premises is possible. AC generator sets may be installed not only to supply emergency power, but also as an in-house electrical base load generator. This can result in a considerable reduction in the cost of electrical power from the regional electricity company and directly provides heat for steam and hot water production, in place of oil/gas/coal fired steam boilers. This utilisation of almost total heat in the fuel oil gives overall efficiencies of up to 90%, compared with efficiencies of 30% for solely electricity generation. For CHP to be economically viable it must operate at high load factors, ideally over 80%, but it is economic for load factors down to 50%. Below the 50% level the savings obtained from steam/water heating do not justify the extra investment for heat exchange equipment.

**4.10** The direct involvement of AC emergency generators in CHP schemes requires considerable equipment retrofit and a change in the philosophy originally accepted for operation in the passive emergency role. A large increase in planned maintenance and operating supervision on engine plant required to operate over say 6000 hours per year instead of 50 hours a year must be expected. In the role as an AC emergency generator set, only limited essential overhaul work is required, but in the role as a base load generator, a long overhaul will become a summer event every second year, particularly with ageing plant. To offset the planned outage of plant, there will be a need for extra or replacement plant, not only to maintain routine supply but also in the event of an engine breakdown during an emergency.

**4.11** When generators are run together in parallel or in synchronism with the regional electricity company supply, more rigorous attention to prospective fault current capacity and the stability of both the health care and personal social services premises and regional electricity company systems must be given.

**4.12** The advice of the regional electricity company must be obtained when CHP or peak lopping operation with parallel generators is planned.

**4.13** The essential power supply must, of course, be guaranteed.

## Location of emergency generating plant

**4.14** Normally the most convenient and economical arrangement will be for emergency generating sets to be centred in a single engine room, probably adjoining the boiler plant. Such an arrangement will reduce plant installation and housing costs and will also be convenient for servicing, routine testing and the provision of a unified bulk fuel supply.

**4.15** On dispersed health care and personal social services premises sites, use of several smaller generating set locations, comprising container or modular housing, will allow reduced lengths of cable installation, costs and sizes. The shorter routes for cables may also improve the flexibility and economic viability of building extensions at a later date.

**4.16**  If there are no planned developments, the overall cost of one central room may be less than having several smaller, separate generating sets located around the site to suit the separate essential services supply load centres.

**4.17**  Normal supply main transformers, switchgear and cables should not be in close proximity to an emergency generator set unless adequate segregation and fire barriers are provided.

## Containerised units

**4.18**  The use of standard ISO type freight containers to house engine-driven generators and gas turbine sets has become accepted practice, for a number of reasons:

a.  the complete unit can easily be lifted, transported and placed in locations that need not be purpose built;

b.  the robust construction allows for a minimum of site foundation preparations;

c.  the whole generating set is enclosed in its own packing case. The container gives good protection mechanically and against the weather;

d.  the structure can be easily insulated acoustically or thermally. On site commissioning can be reduced to a minimum;

e.  for small generating sets the container may incorporate removable covers for plant access, while for sets of 200 kVA or more, there will normally be "walk-in" access.

# Engines and equipment for emergency generator sets

### General

**4.19**  Reliable start-up and continuous operation can be guaranteed provided suitable operational and maintenance routines are followed for engine starting, batteries, engine room, fuel tanks, fuel oil supply pipework, fuel oil, engine jacket coolant and temperatures.

**4.20**  The reliability of starting and running-up times tends to be related to the number of cylinders. It is usually best to select an engine having the most cylinders within the required range of generator output ratings.

**4.21**  Engines of smaller rating have four, six or eight cylinders in line. "Vee" arrangements with 12 or 16 cylinders are used for larger engines of up to 1500kW rating.

**4.22**  For a given cylinder capacity and piston displacement, a higher engine speed will provide a larger power output and will consequently be smaller and cheaper. Higher rates of cylinder and crankshaft wear will be associated with the higher speed, but this is not a significant factor for generating sets used strictly for emergency generation.

**4.23**  For a 50Hz supply, generating sets of the ratings under consideration are available with engine speeds of 1000 revs/min for six-pole generators and 1500 revs/min for four-pole generators.

## Power ratings

**4.24**   Engines should be continuously rated, as defined in BS5514 (ISO 3046). They should be capable of operating at the rated load for a period of 12 consecutive hours, in which time would be included an overload of 10% for a period not exceeding one hour, the prescribed maintenance having been carried out. This is known as a "class A" rating.

**4.25**   Diesel or gas engines should generally be manufactured in accordance with BS5514 (ISO 3046). This standard will be replaced by ISO 8528 under the EEC 1992 harmonisation legislation.

## Engine categories

**4.26**   Four categories of load acceptance are available for various types of engine operation on the basis of percentage load acceptance for the class A rating:

- category 1 – 100% load acceptance

- category 2 –  80% load acceptance

- category 3 –  60% load acceptance

- category 4 –  25% load acceptance.

**4.27**   The load acceptance category will be dependent on the type of aspiration of the engine, that is, naturally aspirated or turbo-charged, and the inertia of the combined engine and generator rotating assembly.

**4.28**   Naturally aspirated engines have a category 1 load acceptance, and are more suited for emergency generation, where the cold start initial step loads required are large in proportion to the engine rating. The frame size of a naturally aspirated engine will be larger than a turbo-charged engine of equal rating.

**4.29**   Turbo-charged diesel engines with category 2 or 3 load acceptance will be more obtainable, efficient and economical for engines having rated outputs above 250kW and operating at high load factors. The cost-effective installation and characteristic must be carefully chosen in consultation with the manufacturer of the generating set.

## Petrol engines

**4.30**   For emergency and continuous generation the petrol engine has no significant advantages over the diesel or gas engine.

## Gas engines

**4.31**   Gas engines of suitable designs have been available since the introduction of the Energy Act 1983 for electrical generation in CHP units. For equal frame sizes and speeds the rated output of a gas engine is 50% – 70% that of a diesel engine. Compression ratios of gas engines are not as high as in diesel engines. Spark ignition is provided to ignite the gas fuel and a carburettor adjusts the air/gas admission ratio, as in a petrol engine.

**4.32**   Turbo-charged, high compression dual fuel oil-gas engines of the diesel type, which use compression ignition, are also available. This type of engine normally starts up on diesel fuel oil, and when running may switch over to dual operation. Compression ignition in the dual operation mode is then achieved by the injection of a small quantity of diesel fuel oil into the engine cylinder containing an air-gas mixture. The air-gas mixture ratio is regulated weak, to prevent pre-ignition in the engine cylinder before the final injection of diesel fuel.

The diesel fuel is approximately 5% of the total energy input in the dual fuel mode.

**4.33**   Due to the zero sulphur, high methane composition of the gas fuel, the gas engine accumulates less engine cylinder deposits and lubricating oil contamination and as a result has longer operating periods at full load between long overhauls, for example over 20,000 hours instead of the 12,000 hours typical of diesel engines.

**4.34**   As a prime mover in the generation of emergency electrical power, the gas engine does not possess as rapid a start-up/load acceptance as the diesel engine. The two main operating disadvantages are:

a.   the greater fire and explosive hazard of gas leakage, in comparison to diesel fuel oil;

b.   the possible non-availability of gas supply at a time of general emergency demand.

**4.35**   In the United Kingdom the natural gas fuel is distributed at low pressure, in general around 20mbar (2kPa), which is only suitable for naturally aspirated engines. Turbo-charged and lean burn engines require gas pressures up to 1.7bar, and therefore to operate a turbo-charged engine it might be required to use a gas pressure booster. When starting in an emergency this becomes difficult as the booster needs to be running before the engine can be started, or alternatively a separate low pressure gas starting changeover fuel system has to be provided. For emergency generation the naturally aspirated engine is the only suitable gas engine.

**4.36**   As the gas supply may not be at sufficient pressure at a time of general emergency, liquid petroleum gas (LPG) in storage tanks is a possible alternative emergency fuel. As a gas it is highly explosive and great care must be exercised with storage and regeneration from liquid. It has a low octane number of 97 (methane 117) and can only be used for naturally aspirated gas engines, which are manufactured up to a maximum of 1000kVA rating.

**4.37**   Gas engines should generally be manufactured in accordance with BS5514 (ISO 3046). British Gas document IM/17 – 'Code of Practice for Natural Gas Fuelled Spark Ignition and Dual-Fuel Engines' gives further reference and guidance.

## Gas turbines

**4.38**   Gas turbines are available over a wide range of ratings for electrical generation, but have several important differences compared with diesel engines.

**4.39**   Gas turbines are not manufactured economically up to 1000kVA rating generally required for essential supply requirements in health care and personal social services premises.

**4.40**   A gas turbine is compact in size for its rating and bed plate construction is simpler. The turbine is rotary and not subject to reciprocating vibration or stresses, and can be installed above ground level.

**4.41**   Gas turbines do not require casing cooling and may be easily installed in waterless areas. Lubricating oil will require forced cooling in air/oil heat exchangers.

**4.42**   Their main disadvantage is the time of one to two minutes required to accelerate the air compressor rotor element to full speed before the generator can be loaded. Step loading a cold gas turbine results in uneven heat expansion in the turbine casing and rotor, which can result in rotational vibration and contact between the fixed and rotating blades. They are more suitable for incremental load increases.

**4.43**    Gas turbines are noisy, especially at the exhaust end. Considerable attenuation of noise level is necessary, involving construction of elaborate ductwork to eliminate the high frequency whistle and pulsation of the exhaust gases. It is essential that they are placed well away from residential and patient areas and are acoustically insulated.

**4.44**    Gas turbines burn natural gas, kerosene, diesel fuel oil or heavy residual oil. Gas compression or heavy residual fuel oil heating and purification may be required before turbine admission. The quality of fuel is most important to ensure a long and fully rated period of operation. Deposits from the metal residues in the heavy residual fuel oil are particularly difficult to remove without special purifying equipment. They build up on the turbine blades as ash, increasing blade to gas friction losses and result progressively in a reduction in the turbine output. The deposits may be removed by low pressure/temperature wet steam admission when the unit is shut down with the rotor turning at barring speed - a process known as "blade washing".

### Diesel engines

**4.45**    Diesel engines will usually be the most effective and economical prime mover for emergency generating sets. For equal engine speeds and frame sizes, the output of a diesel engine is approximately twice that of a gas engine.

### Air cooled diesel or gas engines

**4.46**    For indoor generator sets the use of an air cooled engine is generally restricted to ratings up to 50kW. They are slightly cheaper than water cooled engines, but the cost saving can be offset by the need for increased mechanical ventilation in the engine room due to increased cylinder heat radiating to the surrounding environment. Air cooled engines are noisier as a result of not having an engine water jacket to attenuate the noise generated within the cylinders.

**4.47**    Air cooled engines have the ability to start in low ambient temperatures, but it is desirable that they are equipped with thermostatically controlled, indirect, low temperature electrical heating for the oil sump. This maintains warmth in the engine block and reduces cylinder and bearing wear during cold starts. Trace heating to the fuel oil transfer piping and storage tanks should also be provided.

### Water cooled diesel or gas engines

**4.48**    Water cooled engines have closed-circuit radiator and water jacket cooling systems. The water is circulated by an engine-driven pump to provide the cooling water flow in the radiator. Forced air cooling of the radiator reduces the cooling water temperature before recirculation.

**4.49**    An integral baseplate-mounted air-water cooling radiator is provided as standard by manufacturers for engine driven sets up to about 250kW rating. Above 250kW the engine jacket water cooling radiator would be located in a louvred opening in the outer wall of the building or in a yard outside the building, and be cooled by one or more electrically driven axial fans.

**4.50**   In large engines the engine jacket water cooling and lubricating oil cooling systems both have separate primary circulating pumps. The primary cooling liquids are passed through cylindrical heat exchangers, where they are cooled by secondary cooling water to the optimum inlet temperatures before returning to the engine. The secondary cooling water is cooled in a radiator by fans, or a water-water heat exchanger for heat conservation.

**4.51**   In turbo-charged engines some of the secondary cooling water also flows through a water-air heat exchanger to cool the turbo-charged engine inlet air. In smaller rated engines, air to air radiator cooling of the engine inlet air is possible.

**4.52**   During cold start-ups, to permit rapid heating of the engine block to operational temperature, both the jacket water and lubricating oil are subjected to automatic bypass temperature regulation before the heat exchangers.

**4.53**   With a closed-loop secondary cooling radiator, a water head pressure tank must be provided and should comply with BS417 and be equipped with suitable level indicators and alarms, preferably of the magnetic float type.

**4.54**   The engine jacket and secondary cooling water should be distilled, ion exchange demineralised water or a potable water with minimal chalk or lime solids in solution and be non-acidic, pH 8 to 10.5. The cooling water should be treated with an anti-freeze solution containing a corrosion inhibitor, being either a "universal" type with an ethylene glycol base or one conforming to BS4959. The anti-freeze to water mixture proportions needed will depend on the minimum freezing temperature against which protection is required and on sufficient corrosion inhibitor being present to protect all metal surfaces in contact with the coolant.

**4.55**   A 40% volume (35% minimum to 50% maximum) anti-freeze solution, with suitable multi-purpose additives, should be added to the jacket and radiator cooling waters for full protection in the UK. The advice of the manufacturer should be followed regarding the installation and period of use of an anti-freeze solution.

**4.56**   Under no circumstances should methanol based anti-freeze solutions be used. These have a high evaporation rate at normal engine temperatures and use would lead to rapid loss of coolant.

**Engine starting**

**4.57**   To assist in starting in locations where the engine room temperature is liable to fall below 10°C, thermostatically controlled immersion heaters should be fitted in the engine block at a low level. They should maintain the jacket water temperature at 10°C, in a minimum ambient of 0°C.

**4.58**   In large engines, continuously running, high pressure starting and jacking lubricating oil pumps are required to circulate the sump oil to ease engine starting and run up to speed.

**4.59**   For starting the engine, axial type battery-operated electric starter motors are preferred. On larger engines and in particular those having six or more cylinders it may be necessary to have two electric starter motors to operate simultaneously.

**4.60**   When electric starting is used a means of disconnecting the starter motor automatically after a predetermined interval (say eight seconds) should be used, to prevent heavy discharge of the battery on the first attempt to start. An alarm should be provided to indicate when an attempted start has failed. This should be located in an area that is normally occupied. In some designs a process timing switch is used in the engine control to reinitiate attempted starts at intervals, for example eight seconds' start, with three-second rest intervals. After three consecutive failed attempts, the starter motor is isolated and the fail alarm sounded.

**4.61**   Compressed air starting may be necessary for very large engines, for example 1MW or above. Compressed air starting is more expensive than electric starting within the size range for which electric starting is available. With compressed air starting it is necessary to provide three air compressors – two driven by electric motors and the third driven by petrol or diesel engine to allow for main engine start up difficulties during a black start situation, that is, no electric power supply.

**4.62**   In general, the capacity of air receiver cylinders should be sufficient to give five 10 second turnovers of the main engine at 10 second intervals. Alternatively, a hydraulic mineral oil starting system for engine ratings up to 4MW is also available to replace both electric and air start systems, which would provide a three 12 second starts.

**4.63**   Small static and mobile generating sets, where physically possible, should be provided with hand starting crank levers to replace battery starting in an emergency.

## Engine governors

**4.64**   The engine should be fitted with a speed governor capable of maintaining the rotation speed of the generator within class A governing requirements of BS5514, Part 4 (IEC 3046, Part 4). Shut-down protection to prevent a 10% overspeed shall be provided.

**4.65**   When parallel running of generators is planned a simple mechanical centrifugal governor with a fixed drooping characteristic is inadequate. Hydraulic or electronic (separate or combined) governors with isochronous and/or a maximum 5% variable droop control should be provided to ensure balanced load sharing and stability between generators, according to output rating under changing frequency conditions. This includes a speeder motor controller for biasing the governor control of the generated output in parallel running, or to vary generator revs/min (frequency of generation) when in single running.

**4.66**   Hydraulic governors are slower in response to change, but accurate in operation.

**4.67**   In comparison, electronic governor systems give much finer control and improved transient speed response times for rapidly changing load conditions. The addition of extra electronic circuits leads to:

   a.   a more sensitive load change response time or load differential control, derived from rapid rates of speed change in rotation of the engine/generator rotor;

   b.   automatic synchronising;

   c.   improved sharing of load between differently rated generators, and better load stability, that is safer interaction between generators and also between generators and the regional electricity company supply when operating at power factors close to unity;

d.  electronic governors are more acceptable to the manufacturers of the electronic data equipment and computers for application in a "plant management system" of control.

### Engine failure safeguards

**4.68**  Safeguards should be provided to stop the engine. This should be primarily by de-energisation of the governor DC trip solenoid linked to the fuel rack. This solenoid is de-energised from the master trip relays connected to the electrical protection relays and mechanical protection sensors. Engine shut-down should be triggered by jacket water high temperature, lubricating oil low pressure, emergency and manual stops, and generator protection operation provided by electrical overcurrents, restricted earth fault currents, reverse power currents and loss of excitation field current.

**4.69**  Engine governor overspeed protection should only shut down the engine by mechanical closure of the fuel rack. In such a situation the generator circuit breaker remains closed, to inhibit additional overspeed from loss of load. Opening of the circuit breaker is initiated by secondary protection effects such as reverse power when in parallel as the generator starts to motor, or from low lubricating oil pressure or low voltage/frequency in "island" running as the engine slows down.

**4.70**  Alarm facilities should be provided for monitoring the running condition of the engine and generator, and should include jacket water high temperature, lubricating oil low pressure and high temperature, high exhaust gas temperature, charge air high temperature, fuel oil supply low pressure, and low level cooling water.

### Engine instruments

**4.71**  The following basic instruments are required. They are usually mounted on a panel fixed to the engine or incorporated into a freestanding control panel adjacent to the engine (see paragraph 5.85):

a.  lubricating oil pressure gauge with trip contact or sensor;

b.  lubricating oil temperature gauge;

c.  jacket cooling water temperature gauge with trip contact or sensor;

d.  service hours run counter;

e.  revs/min speed indicator.

**4.72**  Additional electrical protection and overall control for the diesel generator will be required when operating generators in parallel. This advice is given in the Electricity Association (formerly Council) document G 59 – 'Recommendations for the Connection of Private Generating Plant to the Electricity Boards Distribution Systems', June 1985.

## Environmental considerations

### Exhaust pollution

**4.73**  The main gaseous pollutants of fuel oil combustion in a diesel engine or boiler are oxides of sulphur, nitrogen and carbon and the partially combusted hydrocarbons that produce, among other things, the characteristic "diesel odour". Particulate pollutants will be mainly carbon or oxides and traces of polycyclic aromatic hydrocarbons (PAHs).

**4.74**   The amount of polluting effluent from even a small engine unit can be significant and care should always be taken to ensure that it is safely discharged to the atmosphere.

### Sulphur

**4.75**   Oxides of sulphur (SOX) in products of combustion are related to the sulphur content of the fuel, typically 0.3% for light fuel oil to 3.5% for heavy fuel oil. These percentages are expected to be progressively reduced as a result of EEC directives.

**4.76**   The rate of discharge of sulphur dioxide ($SO_2$) for diesel engine exhaust varies typically between 90 parts per million (ppm) for light fuel oil to 1000ppm for heavy fuel oil.

**4.77**   Diesel engines require approximately 100% excess air for combustion, exhausted at approximately 450°C, whereas a boiler requires approximately 20% excess for combustion, and chimney exhausted at approximately 120°C.

**4.78**   The exhaust gas volume from a diesel engine is twice that emitted by a boiler, for the same weight of fuel burned. In boiler plant the concentration of $SO_2$ appears higher, typically 180ppm for light fuel oil to 2100ppm for heavy fuel oil. This equates to the same emission of SOX for the same weight of fuel burned.

**4.79**   Sulphur trioxide ($SO_3$) is also produced in small amounts when the outlet exhaust duct temperature is cooled below the acid dewpoint, approximately 135°C, which then forms sites for the deposition of carbon smuts and the formation of sulphuric acid ($H_2SO_4$), which are discharged as acidic smuts.

**4.80**   The sulphur content in natural gas is nominally zero; there are therefore no SOX components considered as being emitted in the exhaust gases from a gas engine.

**4.81**   The SOX emissions from diesel engines and oil fired boiler plant contribute, in large proportion, to the incidence of acid rain in the atmosphere.

### Nitrogen

**4.82**   Production of oxides of nitrogen (NOX) in a combustion process is influenced by the temperature, pressure and specific features of the combustion process.

**4.83**   Since diesel engines operate at high pressures and ignition temperatures, the production of NOX is approximately six to seven times higher per unit weight of fuel burned than that which occurs in boiler plant. NOX content is not reduced in proportion to smaller loads in a diesel engine as the air intake to the cylinders is constant, that is, it is not restricted by carburettor operation.

**4.84**   In a diesel engine the major NOX product is nitric oxide (NO), and levels vary typically from 400ppm – 700ppm (12.5 – 22.0g/kg fuel). By comparison, in boiler plant the NO concentrations are typically 150ppm – 200ppm (2.1 – 3.2g/kg fuel). The principal remaining constituent of NOX is nitrogen dioxide ($NO_2$) – a more adverse to health and environmentally toxic pollutant. Photochemically produced smog, ozone and acid rain are all associated with NOX emissions.

**4.85**   The concentration of $NO_2$ is typically 5% to 10% of the NO concentration. For diesel engine exhaust gases the maximum $NO_2$ values are about 10% (3.3g/kg fuel) of the total NO emission. In boiler plant $NO_2$ is usually less than 5%, up to 10ppm maximum (0.23g/kg fuel).

**4.86** In comparison, gas engines may operate in the stoichiometric mode at 0% excess air or in the "lean burn" mode at 50% to 80% excess air (the limit of normal spark ignition). The absolute values of NOX can vary considerably depending upon the carburettor setting for air to fuel ratio.

## Carbon monoxide

**4.87** In diesel engines carbon monoxide (CO) content ranges typically from 200ppm – 300ppm (6 – 9g/kg fuel) and from conventional boiler plant around 50ppm (0.7g/kg fuel). For gas engines the CO absolute content can vary with increasing load depending on carburettor setting, as stated for NOX (see paragraph 4.86).

## Hydrocarbons

**4.88** Hydrocarbon products from both diesel and gas engines and boiler plant are equally low, typically between 25ppm – 40ppm (0.6 – 0.9g/kg).

## Air control

**4.89** In gas engines the air admitted into the cylinders is regulated by the throttle valve control in balance with the gas fuel flow through the carburettor. Gas engines are very sensitive to air to fuel ratio changes. Stoichiometric combustion is difficult to maintain without an "$O_2$ in exhaust" feedback control at the fuel input.

**4.90** The volume of air induced into a diesel engine cylinder is constant, at constant speed, irrespective of load variation (see paragraph 4.82).

## Catalytic converters/oxidisers/reducers

**4.91** Catalytic converters have a variety of uses in the removal of undesirable gaseous fumes from the exhaust gases of industrial processes. Chemical conversion of the undesirable pollutant gases to more acceptable gases in the exhaust gas of an internal combustion engine is one of the many uses. This is achieved at temperatures as low as 400°C, exhaust gas temperature, by catalyst chemical agents, usually platinum metal, in the converter unit.

**4.92** No legislation currently exists (1992) for the control of stationary engine exhaust pollutants in the UK. With increasing pollution worldwide this may well change, and may necessitate extensive and expensive retro-fits of catalytic converters to existing plant. However, occupational exposure limits as recommended by the Health and Safety Executive can be enforced for exhaust pollutants in affected work areas (see paragraph 4.106).

**4.93** The John Radcliffe Hospital, Oxford, is currently (1992) the only health care and personal social services premises known to use a first generation two-way catalytic converter for the oxidation of carbon monoxide (CO) and hydrocarbon exhaust gases issuing from a diesel generator unit.

**4.94** The two-way catalytic oxidation units, which are heated by the engine exhaust gases, oxidise the products of partial combustion, mainly hydrocarbons and CO. The principal effect is on the odorous/hydrocarbon component which is almost completely converted to $CO_2$ and steam. Methane when present in the exhaust gas requires a converter unit operating at a temperature in excess of 500°C before the methane oxidation process can start. There is no noticeable effect in oxidising NO to $NO_2$, which is a more toxic pollutant. There is also no change to already oxidised components such as $SO_2$.

**4.95**   The exothermic oxidation of CO and hydrocarbons in the catalytic oxidation unit contributes to increasing the exhaust outlet temperature and provides an additional heat recovery source in the exhaust heat exchanger.

**4.96**   There are two systems for NOX reduction (de-oxidation) to nitrogen (N) and oxygen ($O_2$), namely:

a. three-way non-selective catalytic converter reduction by regulating the air to fuel ratio of the flow through a gas or petrol engine. The carburettor is automatically controlled at all loads to give near stoichiometric combustion of the fuel. This keeps the volumes of exhaust NOX and $O_2$ to a minimum. The CO and hydrocarbon content are oxidised to $CO_2$ and water, and the NOX is reduced to nitrogen and oxygen;

b. selective catalytic reduction of NOX content to N and $O_2$ in large volume exhaust flows by the controlled injection of ammonia ($NH_3$) in a strictly controlled ratio of $NH_3$ to NOX at an exhaust temperature of 300°C – 440°C. Reduction rates of 90% are obtained in the presence of 15% oxygen. This system requires careful management using computer control to maintain the correct balance between $NH_3$ and NOX. It is only suitable for large diesel engine or gas turbine generating units with electrical outputs of at least 1MW. The CO and hydrocarbons in the exhaust gas are oxidised in two-way catalytic units. Sulphur content in the fuel oil should be very low in order not to affect the efficiency and life of the catalyst unit.

**4.97**   Wet scrubber units can dissolve and remove $SO_2$ and $NO_2$ from exhaust gases, but unwanted $SO_3$ can be produced (see paragraph 4.79). They are bulky units and generally not used on small gas or diesel engines. The water vapour produced by the exhaust gas can create a high plume in cold weather.

## Exhaust systems and terminations

**4.98**   The exhaust system should be kept as short as possible and should incorporate a flexible section near the engine outlet manifold to reduce transmission of engine vibration to the remainder of the exhaust system. Bends should have a minimum radius of curvature three times the exhaust pipe diameter.

**4.99**   Whenever practical, it is desirable to fit the exhaust silencer outside the building to reduce the building's internal ambient temperature. When long exhaust piping is necessary (for example, over 7m), its diameter should be increased to minimise exhaust gas pressure drop and the possibility of excessive back pressure reducing engine output. The exhaust pipe should in that case be increased to one-and-a-half times the diameter of the engine exhaust manifold outlet. In such instances an expansion joint will be necessary. The exhaust pipework within the building must be provided with heat insulation and clearance from combustible material to reduce fire risk and protect personnel. Care must be taken when exhaust pipework passes through building fabric to ensure adequate fire safe clearance.

**4.100**   When special silencing is necessary because of the possibility of environmental noise nuisance to adjacent parts of the health care and personal social services premises, it may be convenient to fit an additional silencer, for example one of the absorptive type as well as one of the reactive or baffle types. The manufacturer should always be consulted before long exhaust systems are installed.

**4.101**   Drain valves should be fitted in the lower section of exhaust systems to release any accumulation of water condensed from the exhaust gas.

**4.102**   The position and direction of the exhaust outlets should be chosen to reduce noise levels to personnel and the possibility of recirculating exhaust gases entering buildings through doors, windows or ventilation systems.

**4.103**   Where exhaust gas heat recovery exchanger units are fitted, the exhaust gas temperature at the system terminal can be near dew point, that is 135°C. This can result in a reduction of volume by 40% at the terminal, and an exit velocity below the recommended acceptable level of 20m/sec, especialy at low loads, causing local fallout of odour.

**4.104**   The requirements of the Clean Air Act 1956 apply to the height of exhaust terminations of engines with total heat inputs of 150kW or more per cylinder. The final height of the termination is a function of the total heat input, $SO_2$ content of burned fuel, one of five categories of location, and the height of the building. Guidance is given in the Department of the Environment document – 'Chimney Heights', third edition of the 1956 Clean Air Act Memorandum, obtainable from HMSO.

*Clean Air (Northern Ireland) Order 1981, SI 1981/158 NI4*

**4.105**   Where exhaust gases and odours are found to be an annoyance and/or health hazard, the installation of a catalytic converter may remove the need for a high chimney, and still comply with statutory regulations.

**4.106**   Occupational Exposure Limits (OEL) for $NO_2$, NO, $CO_2$, CO and $SO_2$ are given in the Health and Safety publication Guidance Note EH40/(year of issue), for example E40/91. This publication is used as a criterion to ensure safety within the Health and Safety at Work Act 1974 and other relevant statutory provisions.

*Health and Safety at Work (Northern Ireland) Order 1978, SI 1979/1039 NI 9*

**4.107**   Engine-generated noise must be considered and its long-term effect on the hearing of attendant plant staff. HN(76)126 and the revised appendices to HDN4 (Noise Control) refer.

### Foundations and resilient mountings

**4.108**   The overall weight of the engine and AC generator should be obtained from the manufacturer and a suitable concrete foundation plinth provided that is capable of supporting the generating set and withstanding the engine-generated vibrations, under all required load conditions, and restraining the generator under electrical fault.

**4.109**   Usually sets are provided on a common rigid base plate which will be ready for mounting on to suitable anti-vibration, resilient mountings and which in turn are located between the base plate and foundation plinth. Most manufacturers will supply sets of suitable mountings for their generating sets. In some instances, the base plate consists of a main frame and sub-frame, with the mountings located between the two frames.

**4.110**   Anti-vibration, resilient mountings also help to reduce vibration transmitted into the building structure.

**4.111**   Containerised and mobile generating units should be provided with anti-vibration, resilient mountings to ensure that noise level requirements are met.

## Fuel

### Fuel supply

**4.112**   Diesel engines for emergency generating sets are usually designed for running continuously on light fuel oil that conforms to BS2867 Class A.

**4.113**   The volume of diesel fuel oil stored within the engine room service tank, and arranged for gravity feed of fuel oil to the engine, should be at least 750 litres or equivalent to 10 hours full load running of the generating set. In addition, a fuel oil main reserve for 200 hours full load running for each generator set should be available (see paragraph 4.118).

**4.114**   Gas engines draw natural gas from the British Gas mains supply at a pressure of 20mbar (2kPa) and are subject to agreed tariffs of demand. Gas engines used for emergency generation of essential services supply when the normal electrical supply is lost are dependent on maintenance of the gas supply. If emergency generation requirements occur at a time of heavy demand within the gas supply system, operational difficulties with the gas engine may result, with possible failure of emergency generation. The provision of an additional emergency gas supply, such as an LPG gas storage tank, would be a short-term solution.

### Service tanks

**4.115**   A service tank should be located close to the generating set and preferably installed for gravity feed. Arrangements should be made to ensure oil spillage is not drawn into the generator winding. Tanks should be constructed in accordance to BS799, Part 5 and equipped with the following:

   a.  filler cap and connection for filling, with oil strainer;

   b.  vent to atmosphere by pipe to outside of building;

   c.  dial type oil level indicator, clearly marked to show empty, quarter, half, three-quarter and full. Gauge glasses are not recommended unless fitted with a bottom isolating valve and automatic ball sealing valve;

   d.  low oil level float, overfill and transfer pump running alarms;

   e.  connection for the engine fuel oil injector leak-off return pipe (where necessary);

   f.  drain valves and drain hose connection;

   g.  electrical bond to the generator earth connection. See Figure 7 on page 22.

### Bulk storage

**4.116**   In some instances, for example where an emergency generating set is also to be used for peak lopping of the health care and personal social services premises load (see HTM2011, Emergency electrical services, 'Operational Management', paragraph 3.1) it may be necessary to provide suitably heated and lagged bulk storage facilities with an electrically operated fuel oil transfer pump located close to the bulk storage tank. Figure 7 on page 22 shows a typical arrangement. Where oil is transferred from outside the normal confines of the engine room, lagged trace heating may be required around the oil pipework to maintain the oil fluidity during very cold weather.

**4.117**   Tanks located inside a building in a minimum 5°C ambient environment should be housed in an enclosure of two-hour fire resisting construction. All fuel oil tanks should have a catch pit in accordance with the recommendations of BS5410, Part 3 – 'Fire Protection'. External oil tanks should be located also in accordance with the recommendations of BS5410, Part 3.

**4.118**   Where boilers at the health care and personal social services premises burn diesel fuel oil the emergency generating set may be able to share the same fuel and storage facilities. The capacity of main storage tanks should be sufficient to accommodate the off-loading of a fully loaded fuel oil road tanker together with the fuel oil in reserve. An oil bowser or drum conveyance should be available for site movement of fuel oil.

**4.119**    A hand-operated semi-rotary oil pump should be available for transferring fuel oil from 50 gallon (227 litre) oil drums or other vessels. The hand pump should have a filter fitted with screw caps to prevent ingress of dirt when in storage.

**4.120**    Tanks should incorporate the following facilities:

a.   provision for each tank to be isolated for cleaning and repair;

b.   a vent pipe direct to atmosphere having a cross-sectional area at least equal to that of the filling pipe;

c.   a visible oil level indicator. Where the oil level indication cannot be seen from the filling point, a float operated on another type of overfilling alarm should be fitted;

d.   a lagged tank with steam and/or electric trace heating to maintain the fuel oil at a sufficiently low viscosity may be required during cold weather;

e.   a self-priming fuel oil transfer pump;

f.   tanks, piping and other metalwork associated with the installation should be electrically bonded with the metal base plate of the engine generator set and all earthed in accordance with the IEE wiring Regulations.

## Engine rooms

### Siting

**4.121**    Where practicable, generator rooms should be located at ground level. Basement sites should be avoided because of restricted access for fire-fighting purposes and the possibility of flooding. It is also necessary to consider the fire risk to other parts of the health care and personal social services premises and the effects of noise and vibration on the environment.

### Access

**4.122**    The generator room should be sited so that there is convenient access from a good road to allow an easy approach for fire brigade appliances, maintenance vehicles etc.

### Layout

**4.123**    The layout should be such as to allow adequate space for maintenance. Figures 10 and 11 on pages 25 and 26 show typical arrangements for generator rooms containing one and two generator engine sets respectively.

### Construction

**4.124**    The engine room and any associated room used for oil storage should be of fire-resisting construction; BS6327: 1982 (ISO 6826) – 'Fire Protection of Reciprocating Internal Combustion Engines' and 'Firecode' refer.

**4.125**    The main access should allow sufficient clearance for the passage of both engine and generator. Anchor rings should be provided inside and outside the engine room for drawing in and out the emergency generator set where access overhead is not provided to off-load with an overhead crane.

**4.126**   Doorway openings to other parts of the premises should be fitted with fire-resisting self-closing doors (minimum one hour fire resistance). Doors should always open in the direction of the means of escape.

**4.127**   Floors should have a suitable non-slip and oil resisting finish.

**4.128**   Internal walls should have a finish which, as far as possible, resists build-up of dirt and can be easily and effectively cleaned.

### Ventilation

**4.129**   Adequate ventilation is important to ensure a satisfactory air supply to the engine and to prevent undue temperature rise in the engine room. Where natural gas is used as engine fuel, "gas in air" monitors must be placed in close proximity to the fuel input path and in areas of stagnant air.

**4.130**   An air inlet, which draws directly from the outer air, should have an effective area twice that of the engine radiator cooling system and turbo-charger, if provided, and be positioned in the wall so that the cool air is drawn by the engine radiator cooling fan along the line of the emergency generator set. Where the engine cooling radiator is mounted on the bed plate, air trunking should be used to direct the cooling air through the engine radiator to the outside. This will improve engine cooling and reduce recirculation of radiator hot air around the engine.

**4.131**   With air-cooled engines the building outlet ventilation should have an area at least twice that of the water-cooled engine hot air outlet.

**4.132**   Where natural ventilation is used, outlet vents should be located at high level and above the engine radiator fan hot air exhaust flow. Inlet ventilation openings should be provided at a low level in the engine room where forced ventilation is not used. These areas should be at least 100% greater than those of the outlet vents to allow entry for combustion air.

**4.133**   Ventilation openings should be fitted with louvres or other suitable weather protection and be adjustable to enable them to be closed when the engine is not in use, and so reduce heat losses. With larger sets it may be necessary to have thermostatically controlled inlet and outlet vents, for example motorised louvres controlled by room thermostats with limit switches at the end of each travel direction. With automatic start emergency generator sets all ventilation louvres should be either permanently open or be motorised to move automatically to the open position on engine start-up.

**4.134**   All ventilation ducts should be netted over to prevent entry of birds, vermin or large insects.

### Height

**4.135**   Standard room heights will be satisfactory for small engine generator sets (for example up to 50kW rating), but with larger sets it is desirable to increase the ceiling height to provide work space and space for lifting tackle to assist maintenance. The advice of the manufacturer should be sought.

### Heating

**4.136**   Thermostatically controlled heating should be capable of preventing the temperature in the engine room falling below 10°C. Where electrical heating is used, the heater should be permanently installed and be of a totally enclosed low temperature type. Fuel oil becomes extremely viscous at temperatures below 0°C.

Where fuel routes extend to the outside of the building, pipes should be bound with electric heating tapes or steam tracing and lagged.

## Lighting

**4.137**   An adequate level of illumination should be provided (for example a minimum of 150 lux) in the working area, with good illumination to the front and in the rear of control panels. Modern lamp phosphors may reduce stroboscopic effects in fluorescent tubular lamps. Fluorescent lighting should ideally be used where the luminaires are connected to different phases or be of the high frequency type to avoid stroboscopic effects involved with rotating machines. Regard must be made for the presence of maintained lighting in the engine and control rooms to ensure illumination is present during blackout conditions. Discharge lighting should have a maximum re-strike time of one minute and a run-up time of six minutes and should not be the only source of illumination.

## Socket-outlets

**4.138**   Thirteen ampere, 240V ring or radial main socket-outlets or equivalent should be provided at convenient points around the engine room walls and adjacent to control and electrical relay panels. Residual current devices should be provided for equipment insulation monitoring and to increase personnel safety in areas of increased risk of electric shock.

**4.139**   Provision should be made for 240/110V Class 1 isolating transformers with earthed output winding centre taps, to supply hand tools at 110V, if a 110V ring main supply is not available. Class 2, 240/25V, portable FELV isolating transformers made to BS3535, to supply class 3, 25V handlamps should always be available.

## Fire protection

**4.140**   Where oil or gas engines are used, fire extinguishing protection must be provided over all fuel storage tanks and engines. For gas engines, "gas in air" automatic quick acting isolating valves should be provided in case of gas leakage. For diesel engines in case of a fuel oil fire, a dead-weight isolating valve, closed by a heat fusible linkage, should be provided.

**4.141**   Advice on fire extinguishing equipment should be sought from the local fire authority. Attention should be paid to Firecode.

## Noise

**4.142**   Wherever practicable, generating sets should be located away from patient and staff accommodation areas and other noise and vibration sensitive parts of the health care and personal social services premises and the local environment. All buildings housing engine-driven generator sets should be sound-proofed to conform to the Health and Safety Executive document – 'Code of Practice for Reducing the Exposure of Employed Persons to Noise'. This applies particularly where extended running periods are expected. Where staff are in close proximity and the sound level is at or above 85dB, see HN(76) 126 – 'Noise Control' and Appendices HDN4 (Noise Control), ear-muffs must be provided and worn at all times for hearing protection. In areas where staff are required continuously to supervise and control generating sets running in parallel, a separate sound-proofed annex for electrical control equipment is necessary. See Figure 11 on page 26.

# 5.0 Generators

## Number of generating sets

### Reliability of supply

**5.1**   In small health care and personal social services premises the most economical and convenient arrangement is for a single running diesel emergency generator set to supply power to the essential services. However, for larger premises the best arrangement is to share the load between two or more machines. In such situations the generator sets must be capable of running synchronised and in parallel. Alternatively, a system of two or more generators, with interlocked and interconnected switching, may be necessary to ensure only a single running supply to essential loads.

### Single (island) running

**5.2**   "Island" running represents the simplest arrangement in generator running requirements. Each AC generator will supply electrical power to a discrete isolated system. The equipment connected to the generator in the isolated system will determine the generated load at the fixed engine speed and generator voltage. The total load current and the power factor will result from the total impedance values of all the connected equipments.

### Parallel running

**5.3**   The additional cost of the control and protection equipment required for isolated parallel or mains supply parallel running of AC emergency generators is not normally economically advantageous unless the generating equipment is also used for peak lopping or CHP.

**5.4**   On mains failure, all the duty emergency generating sets start up. The nominated lead set immediately takes up the priority loads, while the second set is automatically trailing and parallels with the first at a selected load demand. When all the generators are operational, the loads will be shared automatically in proportion to governor droop setting and/or generator rating.

**5.5**   Where more than one generator is called upon to provide the essential service load requirements, the load requirement may be sufficiently small that each engine will be running under a proportionally light load condition. Engines should not run at 50% or less load over extended periods. The incoming load can be step-ramp controlled by load discriminating electromechanical or programmable logic controller (PLC) operation to close switchboard contactors or circuit breakers disconnected on loss of supply. The individual step-ramp loads should be regulated to suit the load acceptance category of the engine. If the demand to take up load in a defined time is not realised, or the total load is less than the full load of the lead engine, the trailing engine should be automatically shut down and be relegated to a standby mode. The discrimination of load is generally fixed to the generator ratings. There should be flexibility in the operation of such a system in both the automatic and manual control modes. Facility for the changeover from one to the other of these modes should provide for a smooth power changeover, to avoid erratic load changes to, or inadvertent shut down of, the trailing engine, followed by possible overload and trip of the lead generator. (See paragraph 5.38).

**5.6**   For parallel running, if connected to the regional electricity company mains supply, the windings of AC generators should remain disconnected from earth, as there is an earth connection at the supply transformer, 415V, three-phase winding. When running separately from the mains supply, only the lead or larger generator should be earthed at the neutral switchboard. If one generator is running singly then that generator must be earthed. This single point earthing prevents the circulation of third harmonic currents between the earthed generators and transformer and gives a common earth reference for each phase of the three-phase supply voltage and generator electrical protection. A generator earth is selected automatically or by the operating staff in continuous attendance.

**5.7**   If the regional electricity company supply fails during parallel running, the supply circuit breaker should be tripped by operation of the undervoltage, the rate of change of frequency or the reactive export error detector protection. It is most important in this type of situation that a generator earth is selected automatically or immediately by the operating staff.

**5.8**   Parallel running with the mains supply is only permissible by agreement with the regional electricity company. Reference should be made to the Electricity Association (formerly Council) document G59 – 'Recommendations for the Connection of Private Generating Plant to Electricity Board's Distribution System'.

## Unbalanced loads

**5.9**   It is recommended that the load current in each of the three phases on the emergency generator(s) is balanced to within 20%, to eliminate difficulties in output voltage regulation. Ideally the load should be shared equally between the three phases. The vector summation of any unbalanced phase currents returns by the neutral circuit conductor to the generator that has the stator winding star point switched to earth. For this reason the out of balance current in the neutral should be limited to prevent negative sequence current overloading in the phases of the winding, especially with two or more generators in parallel.

**5.10**   The rated phase current per phase winding should never be exceeded, except under Class A requirements (see paragraph 4.24).

# Generators and voltage control

## Types of emergency AC generator

**5.11**   Generators can be divided into the following main types:

a.   generators with main shaft slip-rings and directly coupled DC exciters;

b.   generators incorporating self-excitation through slip-rings;

c.   generators where the excitation is provided by separate motor-generator sets or from solid-state rectifier units through slip-rings;

d.   generators where the excitation is provided by a directly coupled AC exciter through shaft mounted solid-state diodes, that is a brushless or no slip-ring generator.

## Brushless AC generators

**5.12**   The brushless AC generator is now generally provided for all generating sets, ranging from a few kVA up to high rated outputs, at voltages up to 11kV. The only limitation is the available capacity of diodes mounted on the rotating shaft to pass the required main DC field current to the rotor. The use of slip-rings to provide excitation to generators is now obsolete, except for large power station generators.

**5.13**   The brushless AC generator has no collecting slip-rings on the rotor. The generator field winding DC excitation current is obtained from an AC exciter, mounted at the non-driving end of the generator shaft, either before the non-driving end bearing pedestal or as a stub shaft extension. The AC exciter is wound in a reverse configuration. The stator is the field winding and the rotor the armature winding. Generated AC current is led from the exciter armature winding to six rectifying diodes secured to the generator rotor shaft circumference. The connecting leads may be routed through a hollow section of the generator rotor main shaft from a stub shaft extension assembly, or along the main shaft circumference from an inboard bearing assembly. The AC exciter field winding is provided with regulated current from the automatic voltage regulator (AVR).

### Choice of generator

**5.14**   In an emergency generating set, the only excitation arrangement that is acceptable is the AC generator with brushless excitation. Brushless AC generators have been in service for many years and have proved reliable. Unless otherwise specified a brushless machine should be supplied.

**5.15**   The AC generator should comply with BS5000, Part 3, 1980 to match the engine Class A output rating. A Class B stator and rotor winding insulation temperature rise is sufficient for UK ambient temperatures. A standard design using Class F insulation is acceptable if economically advantageous.

### Rated output

**5.16**   The AC generator should be designed to supply its rated output continuously at a power factor of 0.8 lagging, nominated voltage and frequency at the engine revs/minute.

### Automatic voltage regulators

**5.17**   The automatic voltage regulator (AVR) is the central feature of the generator output control and stabilising reference for the electrical protection relays and metering. It is an amplifying control device which responds to an input of variable and fixed voltage references, derived from:

    a.  a generator terminal voltage transformer;

    b.  a voltage element in proportion to the load current;

    c.  the manual trim controller;

    d.  fixed reference voltages based on the generator rating.

The AVR should operate normally in the auto mode, items (a) - (d), but should have selective control to operate in the manual mode, items (c) and (d), in an emergency.

**5.18**   The output power of the AVR amplifier is derived from the power source listed in paragraph 5.11. In the case of the brushless generator, the AVR fixed voltage references are set to give a no-load rated terminal voltage. These are set by the manufacturer according to the operating instructions document.

**5.19**   The variable feedback control voltages are extremely sensitive to generator voltage and load current flow changes and can force a rapid change in the generator terminal voltage to correct an excursion from the fixed reference voltage. Any system or generator source voltage transient effects are fed into the AVR and amplified as AC exciter DC field current in opposing magnitude to correct the voltage transient changes back to the set reference values.

**5.20**  The output current from the AVR provides the DC field current to the AC exciter. The main field DC excitation current is derived from the AC exciter, after passing through the rectification diodes, and finally controls the generator terminal voltage, in relation to the load demands and any system voltage changes.

**5.21**  Certain transient values can be detected in the AC exciter DC field current by oscilloscope and used as diagnostic monitoring to detect developing faults in the field control system, for example faulty rectifying diodes or rotor earth faults.

**5.22**  The AVR output current, or AC exciter DC field current, is in strict ratio to the main field DC current and that relationship can be metered to represent the main field DC excitation current.

**5.23**  The regulated current to power the AC exciter field winding is obtained from the AVR by one of three methods:

a. by the resulting generated terminal voltage feedback to the AVR through a small power transformer connected at the generator output terminals, that is, self-excited. This voltage is not self-sustaining under system fault. A voltage maintenance element should be provided for protection relay discrimination in the sub-circuits;

b. by direct feed to the AVR from a central battery source, that is battery excited. A stable voltage under system faults;

c. by a permanent magnet pilot exciter mounted preferably on an extended stub shaft or alternatively driven by belt from the main shaft. This pilot exciter provides a constant voltage output to the AVR at the synchronous speed of the generator, that is pilot excited. This is the preferred method of generator excitation.

**5.24**  Solid-state electronic AVRs are now the only type used. They are reliable and more sensitive to changing conditions at the AC generator output terminals. This gives better overall load control and dynamic stability for generators running in parallel or sharing load when connected to the regional electricity company supply.

**Dual channels**

**5.25**  If it is desirable to have fully automatic AVR operational protection, then dual "auto" regulator channels should be considered. A more economically advantageous installation would be to utilise a manual channel as the back-up to an auto channel, which operates typically as follows:- An equal regulator setting for generator voltage is maintained continuously by an "auto follow-up" slave control circuit in the manual channel. Any unbalance between the two regulator channels is shown in an "auto/manual" excitation balance meter. When small this unbalance between auto and manual can be corrected by the manual trim controller, before an operational changeover to auto/manual or vice versa is required. In the event of a failure in the auto channel, the control will trip and alarm automatically from auto to manual channel mode. In the event of failure in the manual channel "auto follow-up", any unbalance will trigger an alarm.

**5.26**  In the manual channel mode, the generator voltage control is no longer responsive to changing generator load or voltage in the system. An operator will be required to regulate continuously the generated voltage on change of load or system voltage.

## Voltage regulation limits – grade 2

**5.27**   The generator terminal voltage should be automatically maintained within ± 2.5% of the rated voltage at any load, from no load to 10% overload, between unity and 0.8 power factor (lagging). The droop in the AVR-controlled voltage should be substantially the same for all generators.

**5.28**   After a change in load from zero to 60% the voltage should not vary by more than ±15% of the rated voltage and should return to within ±3% within 0.5 seconds. On starting, the voltage overshoot should not exceed 15% and should return to within 3% in not more than 0.15 seconds. The manual trim controller should be able to adjust the AVR-controlled generator terminal voltage to within ±15% of nominal limits.

## Excitation and load control

**5.29**   An AC generator running singly at a constant governor-controlled speed and unconnected to the regional electricity company supply will have the current and power factor, at the rated voltage, determined by the impedance characteristic of the load.

**5.30**   When AC generators are running in parallel, whether connected to the regional electricity company supply or not, a continuous interaction occurs between the governors and AVRs of AC generators, which, if not regulated, may eventually lead to dynamic instability in a generator operating at or near rated output.

**5.31**   Excitation field current can be increased or reduced by manual adjustment in the manual trim controller, even in auto control. In parallel operation with other generators and to a greater extent with the regional electricity company supply, field current adjustment not only varies the voltage generated at the generator switchboard, but also causes a redistribution of inductive reactive power load between the other parallel generators. This results in changes to the load currents and the power factors of each generator and the supply transformer, that is, the incoming supply. Any small changes in the field current by a generator manual trim controller should be counterbalanced by small opposite adjustments in the manual trim controllers of the other parallel AC generators. Excessive reduction in field current can cause instability and loss of synchronism in the AC generator near or at full load. Likewise, exceeding the maximum field current will cause generator rotor overheating and will also reduce the stability margins in adjacent parallel generators.

**5.32**   In a system of balanced governor control and parallel running generators, improved stability and load sharing can be obtained by an additional input element of AVR auto-control which responds to the summation of the reactive or quadrature vector components in each generator load current. This unified control mode, known as the single/parallel mode, is individually selective at each generator control panel.

## Governor control

**5.33**   The scope of a governor to regulate the supply of fuel to an engine, hence the output from an AC generator, is determined by the increment of additional movement provided by the governor to open or close the fuel rack, at the governor set synchronous speed. Any outside cause to change the generator speed or any speeder motor adjustment applied to the governor setting, is equivalent to a change in electrical demand or output required.

**5.34**  The governor overall control and speeder motor bias control function jointly in the speed domain. For single "island" running, they automatically maintain the speed at the selected load, within a speed range of 100% to 95% and a load range of no-load to full-load respectively, that is the governor speed characteristic. For single running, a speeder motor auto control may be provided with isochronous speed control to maintain constant speed, hence constant frequency of supply with changing load demand. In parallel running, as described above for single running, the governor and speeder motor controls function jointly. The governor is now tied to the fluctuation of the system frequency, at a selected generator output.

**5.35**  A rise in system frequency will initiate the governor to close the engine fuel rack, to reduce the generator output, likewise a fall in system frequency will further open the fuel rack, if it is not already fully open, and attempt to increase the generated output. This functions within the range of the governor speed characteristic, over the speed range of 100% to 105% synchronous speed, giving a load change from full-load to no-load respectively.

**5.36**  From no-load to 110%, the governor should be stable and sensitive, and respond to prevent speed excursions reaching 110% speed. At a speed of 110%, the governor overspeed protection operates to close the engine fuel rack, cutting off the fuel supply to the engine.

**5.37**  Control of running generating plant should only be exercised by adequately trained authorised or competent persons.

## Power management systems

**5.38**  Unified electronic AVR and governor control systems can lead to the development of complete power management systems (PMS) of logical auto control. These systems can be extended to multi-generator parallel operation, programmed by computer, to selectively start and synchronise, vary the load, shut down, and monitor for operational records and breakdown incidents, without operator attendance (see paragraph 5.3). A mode of operation such as PMS is complex and outside the scope of a simple AC emergency generator installation, but is proposed as a possible arrangement for the control of more complex and ambitious installations.

## Power factor correction

**5.39**  Power factor correction (PFC) units have capacitive reactive power outputs and are installed to compensate (negate) excessive inductive reactive power demand. These may overcompensate when in circuit with generating plant. The net result of uncontrolled overcompensation will be seen as a higher than normal supply voltage and leading power factor. The PFC should be isolated from circuit when not required, unless it is reliably and automatically controlled.

**5.40**  A generator AVR will try to reduce the supply voltage, by reduction in the generated inductive reactive power,  to reduce any PFC overcompensation. The indicated power factor will increase towards unity, or to a leading power factor. This change in power factor indicates a generator is operating in a less stable operating region.

**5.41**  In situations where PFC overcompensation may occur, audible alarms to indicate generator minimum field excitation current should be provided. The additional option of automatic field excitation current "forcing" at the minimum field excitation current to maintain generator synchronism, should be provided on larger generators.

**5.42**   Where static rectifiers represent a large proportion of the generator load (see paragraph 3.83), it may be necessary for the capacitor to be de-tuned, by means of a small series inductor, to outside the resonant frequency range of harmonic voltages injected into the supply by a rectifier without input filters.

## Waveform

**5.43**   The voltage waveform of the AC generator should be sine wave in shape and within the permitted limits of BS4999, Part 40. This requirement applies to all AC generators, regardless of output rating.

## Terminals and cables

**5.44**   The main, auxiliary and earthing terminals of AC generators should be adequate in size for the designed current to be carried, and be properly secured, mounted and protected. Means should be provided for preventing the terminals from working loose after the connections have been made. Terminal boxes should be of steel construction suitable for armoured cable. Terminal boxes should comply with BS4999, Part 145, 1987 and terminal markings to the Institution of Electrical Engineers document – 'Regulations for Electrical Installations'. Explosion diaphragms and moisture absorbers should be provided in the phase and neutral terminal boxes of HV generators.

**5.45**   Single core cables with aluminium armour should enter gland plates at insulated glands. Gland plates for single core cable entry should be of the non-magnetic metal or slotted and brazed steel type to prevent gland plate circulating currents, if total cable currents are in excess of 500 amperes.

**5.46**   The non-driving end bearing pedestal of the generator must be electrically insulated from direct metal contact with any part of the main unit equipment or base plate. This is a precaution to prevent induced voltages in the rotor shaft producing circulating currents through the engine and generator main bearings which would cause electrolytic decomposition of the bearing surfaces.

**5.47**   AC generator cables and glands should be of adequate load and fault current rating suitable to withstand the largest prospective fault current for the most onerous connection between generators and regional electricity company supply. All directly connected switchboard busbars and switching devices should be full load and prospective fault current rated.

## Overcurrent protection

**5.48**   Generator protection should be able to discriminate between a sub-circuit or a sustained self-destructive fault current. A sustained current of three times full load should be cleared within 10 seconds. Fuse protection will be generally too slow to operate. For earth fault currents, the maximum value of the generator 415V earth fault loop impedance must be considered. Restricted earth fault (REF) protection may be used and balanced to detect winding to stator core faults.

**5.49**   Magnetic overcurrent limiting devices, such as 415V, three-phase moulded case circuit breakers (MCCB), are preferable. The MCCB current/time minimum setting to trip must be **less than** the generator transient current/time characteristic curve value to minimise stator insulation heat damage for a sustained self-destructive fault current.

# Controls, changeover contactors, relays etc

## Changeover contactors

**5.50**   The range of automatic changeover switches is wide. Sophisticated automatic changeover switches are available in excess of 2500 amperes rating.

**5.51**   For emergency AC generator requirements it is essential that the rating of the main automatic changeover switch is adequate to accept single (island) generator or supply the system high surge currents as applicable.

**5.52**   The total current for heating, lighting (tungsten and fluorescent) and electric motor rated full load currents should be added together for total main switch rating. It is advisable that industrial power, small power and lighting currents to sub-circuit automatic changeover switches should not be mixed.

**5.53**   Changeover switches should be housed in dust proof enclosures.

**5.54**   BS764 – 'Automatic Change-Over Contactors for Emergency Lighting Systems', the manufacture of double pole changeover switches up to 100 amperes rating and BS5266, Part 3 – 'Small Power Relays for Emergency Lighting Applications up to and Including 32 Amperes Rating' refer.

**5.55**   An important feature of emergency lighting automatic changeover switches is that they are designed to "fail safe". The most simple design is when unlatched contacts are held closed electromagnetically. On failure of the normal supply, the mechanically linked contacts are changed over by the action of a charged return spring or by gravity to the emergency supply mode. This design will operate rapidly, that is will transfer to an emergency services supply in the event of loss of normal supply or failure of the control sensors, and retransfer on resumption of normal supply.

**5.56**   Essential mixed loads of tungsten and fluorescent lighting or power starting currents should be controlled in the essential supply by sub-circuit automatic timer switches. The auto starting of electric motors should be controlled by auto-timer operation to, or in, the motor starter panels.

**5.57**   Manually operated sub-circuit changeover isolator switches can be used where operators are available to change over between supplies not on a priority rating. These manual changeover isolator switches should be clearly identified for rapid operation in emergencies. Manual changeover switches are intended only for use as load switches.

**5.58**   The main auto-changeover switch is normally required to withstand the prospective through fault current at the main switchboard, but not interrupt that current. Recent developments can specify for auto-changeover switches with circuit breaker characteristics.

**5.59**   Some main auto-changeover switch arrangements used with AC emergency generators may employ two separate contactors or circuit breakers of the draw-out pattern. They each have their respective tripping and closing coils controlled by the normal or emergency supplies, in conjunction with voltage/frequency control relays, time delay and a mechanical interlock to prevent the possibility of the two sources of supply being connected together

**5.60**   The auto-changeover switch arrangements using circuit breakers should comply generally with BS4752, Part 1, 1977 – 'Switchgear, Control Gear, Voltages up to 1000V AC'. Contactors should comply generally with BS5424, Part 1, 1977 – 'Contactors, Control Gear for Voltages up to and Including 1000V AC and

*An IEC specification draft document No IEC 947-6-1 – 'Low Voltage Switch Gear and Control Gear: Automatic Transfer Switches' is awaiting IEC harmonisation (1992). This document specifies for dedicated automatic transfer switches having either load switch or circuit breaker operating characteristics. IEC 947 is a composite multi-part standard embracing most electrical switching devices built into motor control centres and switchboards. This will supersede the equivalent British Standard.*

1200V DC'. A switch should be capable of satisfying the making and breaking capacity test with inductive load having a 0.35 power factor lagging, and be stable for prospective maximum through-fault currents. (See "References" paragraph 2.62.)

**5.61** Contactors that are subject to the 415V system prospective fault level currents should be mechanically latched in both the normal and emergency closed positions. The geometrical construction of circuit breakers and contactors assists the main contacts to open rapidly when subjected to large through fault currents. The opposing reaction between electromagnetic forces produced at, and around, the main contacts assists in driving the main contacts apart.

### Bypass switches

**5.62** Where operationally justified and economically advantageous, the auto-changeover switch should incorporate a bypass facility to permit maintenance/testing while connected to normal or emergency supplies. The bypass will permit full routine tests at either generator no-load, essential load or at increasing loads up to full load with a connected ballast resistor load. The bypass switch will allow, concurrently, maintenance and tests on the auto-changeover switch without disturbance or interruption to the system load (see Figure 9 on page 24).

### Transfer times

**5.63** Rapid changeover action is a requirement for all auto-changeover switches, to provide an essential supply to life-maintaining facilities. An emergency generator will be required to provide a steady-state supply within approximately 15 seconds from "start" initiation.

**5.64** Voltage sensitive relays for initiating change of operating condition, either for sequence starting or stopping the prime mover, should have voltage setting ranges of 75% to 95%, normally set at 85% and able to discriminate a frequency variation of ± 5%, set at 95%Hz.

**5.65** The time delay devices incorporated with the voltage sensing should have adjustable range time settings. For initial start of an engine a range of 0.5 to 6 seconds is advised with a setting of say 3.0 seconds, depending upon the regional electricity company auto reclose scheme of protection for their system circuit breakers. This is followed by a transfer to the emergency supply after establishing correct emergency generator electrical conditions with a range of 0 to 1 minute and a setting of say 1.0 second.

**5.66** The choice of a low operating voltage for the voltage sensitive relays controlling the auto-changeover switch will not in itself be a complete safeguard against spurious operation, since transient loss in voltage may fall below the minimum hold-on operating voltage of the relay. In addition to the short time delay necessary to expire before initiating a prime mover start signal, it is also necessary for the main auto-changeover switch to remain closed to the normal supply. This is effected by a mechanical latch until after such time as the AC emergency generator has established the correct essential supply voltage and frequency, and the generator circuit breaker has closed.

### Retransfer of essential supply circuits from emergency generator supply to normal supply

**5.67** With a secure essential supply from the AC emergency generator the need for rapid reconnection to the normal supply is not great. Manual reconnection to normal supply is preferable when suitable manpower is available. Use of manual switching will also ensure that the AC emergency generator is given the opportunity for a longer run on load than may be offered by routine testing.

**5.68**   As an alternative to manual switching, automatic retransfer from emergency to normal supply may be achieved by the use of suitable voltage sensing relays connected into the normal supply (see paragraph 5.64).

**5.69**   Time delay devices should be incorporated with the voltage sensitive relays. Before an automatic rechangeover to normal supply is initiated a delay timer of 0-30 minutes' range, with a setting of, say, 20 minutes, should interpose to ensure that the normal supply is secure.

**5.70**   It is essential that after rechangeover the engine should be allowed to run unloaded to lower the temperatures in the piston cylinders and fuel injectors before an engine manual shutdown is initiated. For an optional automatic shutdown, a delay timer of 0-6 minutes' range with a setting of five minutes should interpose following the rechangeover operation to normal supply before an engine shutdown is automatically initiated.

**5.71**   At any time when the normal supply is reinstated the "break before make" changeover from AC emergency generator to normal supply is an almost instantaneous switching operation. At this time, damage to large electric motors and starters of around 37kW rating or more can occur owing to the out-of-phase relationship between the motor's back EMF and that of the "making" normal supply. Double voltage levels can be imposed with heavy surge currents into all motor stator winding and starter contactors before the two voltages are permitted by the rotating inertia to pull into synchronism.

**5.72**   The retransfer time delay should be sufficient to allow a decay in all electric motor back EMFs to a value of 25%, that is, a total possible 125% voltage is considered the worst condition of reconnection to normal supply. The time delay should not exceed 15 seconds. The exponential time constant of the larger motor generated voltage decay should be obtainable from the manufacturer. This time delay may be one or two exponential time constants, or four to eight seconds to 25% voltage.

**5.73**   During a retransfer when out of phase voltages can result in surge inrush current damage and where there is no automatic synchronisation facility, the following action is to be considered:

   a.  a health care and personal social services premises system shut-down at the generator circuit breaker. This results in an immediate electric motor no-volt trip and a subsequent manual or auto-timer restart after a delayed manual normal supply reconnection at the main auto-changeover switch;

   b.  a delayed retransfer, using a main auto-changeover switch with a neutral "off" position. The loss of emergency supply trips all the closed electric motor starters on no-volts. When the normal supply is reconnected all tripped motors restart by auto-timer.

**5.74**   In unbalanced systems, transient voltage disturbances can result during the retransfer of neutral connections from the generator earth system to the transformer earth system, where no "make before break" neutral-earth contacts exist (see paragraph 5.8).

**5.75**   When a UPS inverter is supplied normally from the essential services supply, a rechangeover from emergency to normal supply does not introduce phase voltage transients in the inverter, where the DC supply acts as an isolation barrier between the AC input and load.

**5.76**   If the load provided by a UPS operating from battery supply only, is switched over to a mains supply bypass operation, that switching operation should be delayed until the UPS output frequency voltage and phase difference are auto-aligned with the bypass input mains power supplies.

### Relays

**5.77**   The provision of satisfactory and reliable control relays is an important part of emergency supply equipment. Compliance with British Standards, harmonised IEC specifications or British Telecom specifications for relays should be specified where applicable.

**5.78**   The control relays will be operated only as frequently as the regular no-load test start-up and the monthly run on the available esssential load. It is essential that relays should be of the best quality available. Relays should preferably be of the gold-plated contact type and fully AC or DC rated to withstand the arcing involved in making/breaking. They should be dust tight with plug-in bases and have standardised contact arrangements to allow simple replacement and stock holding. All relays should be nominated and relevant information fully tabulated on the circuit diagrams for relay coils and contacts identification. (BS5992 refers.)

**5.79**   The electrical protection relays should be of the highest quality. For simple arrangements, an electromechanical relay is suitable. For more complex arrangements, the more proven makes and designs of electromechanical and/or electronic (static) relays should be chosen.

**5.80**   Current transformer ratios should  be carefully chosen to operate with good current linearity for metering up to and beyond the normal full load currents to the permissible overload, or for protection relays up to short-circuit currents. Protection relay and current transformer values should be simulated in the relay secondary injection test equipment available from specialist manufacturers.

**5.81**   Where high frequency current injection techniques are used for essential services switching during emergencies, the electromagnetic compatibility (EMC) of the equipment with health care and personal social services premises services and medical equipment must as a legal requirement be considered.

## Wiring, controls, panels and cubicles

### Wiring

**5.82**   Interconnecting wiring between control and indication equipment mounted on the diesel generator set main framework should have adequate protection against mechanical and oil damage, for example wiring enclosed in screwed steel conduit or oil resistant cable. Flexible metal conduit and flexible wiring which provide a similar protection against mechanical damage should be used between rigid components and those that are subject to vibration whilst the set is running. MIMS cable should not be used as the metal sheath is subject to vibration fatigue failure.

**5.83**   Control panel wiring should be in PVC insulated stranded conductors, complying with BS6231. The minimum size of conductor should be 1.5mm$^2$. All wiring should be adquately supported and protected. Panel wiring looms should be looped where wiring is subject to movement at hinges of doors and pull-out trays. Wiring at different levels and types of voltages should be segregated and colour coded to the IEE Regulations. All termination blocks should be identified and voltage segregated. All terminal wires should be single terminated using linked blocks and marked with secure identification ferrules, as tabulated in the cabling schedule and recorded on the circuit diagrams. Spring loaded terminations (class 1) should be considered for the generator main electrical protection.

**5.84**  Terminating lengths of incoming control wires should be generous and loomed to allow subsequent reterminating at the remotest terminal. Incoming, interconnecting control cables should be wire armoured and screened overall, with wires seven-stranded, 2.5mm$^2$ size. For 50V rated alarm and indication circuits, the wiring should be 1.0mm$^2$ twisted pair type or individually screened and earthed.

## Controls

**5.85**  With small-and medium-sized generator sets it is usual to mount the instrument control panel on the main frame of the AC generator. With large sets, a free-standing control cubicle is usually necessary.

**5.86**  Where control panels are fixed to the generator frame or bedplate, resilient mountings should be used at the bolted fixing points to reduce vibration fatigue in the instrumentation.

**5.87**  When parallel running is required a further free-standing control cubicle will be required to accommodate the additional controls to synchronise the AC generator and to accommodate electrical protection relays.

**5.88**  Side and/or rear access panels should be provided as necessary to control cubicles. All removable panels should be secured to the cubicle frame by set screws and be removable only with maintenance tools.

**5.89**  All panels, both internal and external, that can be removed for access to live components that operate at potentials of 50V or greater should be marked "DANGER . . . VOLTS". Danger notices should comply with BS2754.

**5.90**  It should be specified whether top and/or bottom cable entry to the cubicle is required and due allowance made for cable minimum radius of bending. Suitable provision should be made for tool access above and below the cable glands. The cubicle, trunking or conduit enclosing the wiring should be adequately protected to prevent entry by vermin.

### Typical components on control panels and generating units

**5.91**  The installation of components listed will be subject to the ratings and project requirements:

    a.   interlocked auto-changeover switch;

    b.   time delay mains failure relay for engine start and auto-changeover;

    c.   voltage/frequency sensitive relay for AC generator circuit breaker;

    d.   AC generator overcurrent relay;

    e.   time delay relay for automatic normal supply reinstatements;

    f.   time delay relay for engine cooling;

    g.   bypass switch to auto-changeover switch for test and maintenance;

    h.   push button – manual start engine;

    j.   push button – emergency stop (red with yellow backplate);

    k.   auto-start sequence switch;

    m.   mains failure simulation switch or push button with reset spring;

n.   fail to start relay with audible/visual warning;

p.   lubricating oil pressure gauge with engine low pressure alarm;

q.   engine high temperature relay with audible/visual warning;

r.   generator voltmeter with phase selection rotary switch;

s.   generator ammeter with phase selection rotary switch;

t.   frequency meter;

u.   battery charger and batteries for start control, alarm and indication;

w.   battery and battery charger isolation and visual indicator;

y.   battery ammeter;

z.   battery voltmeter;

aa.  battery reverse current cut-out;

bb.  hours run meter;

cc.  indicator lamp - mains supply available;

dd.  mains voltmeter;

ee.  automatic voltage regulator;

ff.  voltage manual regulator;

gg.  engine sump heater switch;

hh.  kW hour integrating meter;

jj.  kW indicating meter (three-phase, four-wire);

kk.  power factor meter;

mm.  generator circuit breaker with indicators and protection;

nn.  failure to start engine alarm;

pp.  generator on-load indication (circuit breaker closed);

qq.  generator off-load indication (circuit breaker open);

rr.  overcrank (fail to start alarm).

## Generators in parallel or connected to the regional electricity company supply

**5.92**   the following additional equipment is required:

a.   automatic synchronising unit;

b.   synchronising indicator;

c.   governor controller (speed/load control);

d.   excitation controller (voltage, power factor);

e.   circuit breaker controller and auto/manual selector;

f.   power factor meter;

g.   field current meter;

h.   electrical relay panel with protection to Electricity Association requirements;

j.   generator circuit breaker with trip circuit failure alarm (high voltage);

k.   single/parallel mode selector switch;

m.   auto/manual excitation control changeover switch;

73

n.  auto/manual excitation balance meter;

p.  earthing switch controller;

q.  earth indication – switch closed;

r.  neutral earth switchboard;

s.  lamp test facility for all alarms and indications.

**Warning notices**

**5.93**  All permanently fixed signs installed with the intention of being health or safety messages should have a geometric shape, colour and pictorial symbol conforming to BS5378, Part 1, to ensure compliance with the Safety Signs Regulations 1980, No 1471.

*Safety Signs Regulations (Northern Ireland), SR 1981 No 352*

**5.94**  Conspicuous safety notices should be provided in a prominent position on each side of the emergency AC generator set to read:

| |
|---|
| DANGER |
| THIS MACHINE IS AUTOMATICALLY CONTROLLED DO NOT WORK ON IT UNTIL THE STARTING AND FUEL SUPPLY MECHANISMS ARE ISOLATED OR DISCONNECTED |
| WEAR EAR PROTECTORS WHILE ENGINE IS RUNNING |

**5.95**  A notice should be placed near the auto-changeover switch warning against leaving the generator isolated at the bypass switch or circuit breaker.

# 6.0 Batteries for emergency supply

## General

**6.1**  Batteries for emergency purposes may be of the nickel-cadmium (alkaline) or lead-acid type. See Figure 6 on page 21.

**6.2**  Lead-acid batteries are less expensive than alkaline batteries, but are bulkier and heavier.

## Alkaline batteries

**6.3**  Alkaline batteries should conform to BS6260 or BS6115, as applicable, and be operated to BS6132, 1983 – 'Code of Practice for Safe Operation of Alkaline Cells'.

**6.4**  The alkaline battery has the advantage of being more robust than the lead-acid battery. It is also almost non-self-discharging and can accept heavy recharge and discharge currents. An operational life of more than 20 years can be expected. Capacity may not be maintained unless deep discharge/recharge is regularly carried out (see HTM 2011, Emergency electrical services, 'Operational Management' paragraph 2.28).

**6.5**  The alkaline battery is often used to operate UPS, circuit breaker and engine control panel controls and indicators, where a short routine maintenance may be difficult.

**6.6**  Dry sealed alkaline cells found in self-contained emergency lighting systems have restricted lives. Five years operational life may be expected under reasonable circumstances.

**6.7**  The nominal discharge voltage per cell (1.2V) is less than for the lead-acid cell and consequently more cells per battery are required. It also has a higher internal resistance than lead-acid cells. Specific gravity remains constant at 1.18.

**6.8**  Both plates in the cell are formed from perforated steel strips. The positive plate contains pockets of nickel hydroxide and the negative plate contains pockets of cadmium hydroxide. The plates are bolted to the group bars. They do not shed or disintegrate with use.

**6.9**  Health regulations, due to the cadmium, require the manufacture and disposal of nickel-cadmium batteries to be carefully managed.

## Lead-acid batteries

**6.10**  There are a range of lead-acid batteries on the market. Mainly:

  a.  Plante;

  b.  open-vented high performance flat plate;

  c.  open-vented plate automotive – not recommended for emergency use;

  d.  open-vented tubular plate – no British Standard available;

  e.  sealed recombination – in either high performance or automotive ranges.

**6.11**   The lead-acid battery should conform to BS6290, Parts 1 to 4, and be operated to BS6133, 1985 - 'Code of Practice for Safe Operation of Lead-Acid Cells'.

**6.12**   The cell nominal discharge voltage is higher (2.0V) and the internal resistance lower than for the alkaline cell. Their main disadvantages are lack of mechanical strength in handling and poor resistance to vibration.

**6.13**   They are more self-discharging and their specific gravity varies with low to high charge (1.19 to 1.22). They therefore require closer maintenance  supervision than alkaline batteries. The plates also shed and disintegrate with use.

**6.14**   The Plante battery is supreme and the most expensive of the lead-acid range. It is more reliable for engine starting, emergency lighting and switchgear duties. The positive and negative plates are made from pure cast lead of lattice construction and are cast to the group bars. The negative plate is impregnated with lead oxide paste. An operational life of 20 years can be expected.

**6.15**   The plates in the high performance "flat plate" battery are both of lead alloy. The battery is cheaper than the Plante and an operating life of 12 years can be expected. It is heavier and more bulky than the alkaline battery. It is suitable for emergency duties when purpose-made for that function.

**6.16**   The "automotive" battery is essentially an economy flat plate battery with an expected operating life of no more than two years. It is not suitable for any emergency functions or long-term duties involving continuous trickle charge to maintain a charged state of standby.

**6.17**   The "tubular plate" battery is of different construction to the flat plate. The positive plate consists of lead alloy spines surrounded by synthetic fibre tubes filled with a mixture of lead oxides. The negative plate is of the pasted lattice type. The battery is manufactured in the UK to the German Specification DIN 40736. The tubular plate battery is ideal in applications that require frequent charge/discharge cycles. When on emergency duty an operational life of 12 years is expected, but on electric truck duty this reduces to only five years.

**6.18**   The "sealed recombination" battery is used extensively for emergency and UPS requirements. The positive and negative plates are separated by a porous micro glass-fibre spacer which absorbs the electrolyte and assists the diffusion of oxygen towards the negative plate for reaction with the hydrogen to form water. It requires the minimum of maintenance; no topping up with distilled water is required. The volume of electrolyte is very small in comparison to the open-vented type battery.

**6.19**   A sealed recombination battery container is not required to be transparent or translucent. For the larger lead-acid open vented battery which requires water additions, transparent or translucent jars should be provided.

**6.20**   To be at their most effective, batteries should be kept at a warm ambient temperature, ideally around 20°C.

## Aluminium-air batteries

**6.21**   An aluminium-air primary battery has been developed with a continuously rated output of 600 or 1200 watts at 24 or 48V DC, for a minimum period of 48 hours. This period can extend up to 80 hours, depending on the load demand on the rated capacity.

**6.22**  The battery is suitable where a small extended power is required to support a lead-acid or alkaline battery emergency system of supply. It may be economically advantageous where a small rated diesel generator would be required to provide supply for a short period. The response to demand is not immediate, as it requires approximately five minutes to reach full rated output. Once established, the auxiliary equipment electrical power required to maintain the electrolytic reaction is self-supporting. The battery operation is fully automated and monitored by instrumentation.

**6.23**  On full depletion, the aluminium plates and electrolyte of aluminium hydroxide solution will require replacement.

**6.24**  The aparatus is environmentally compatible and emits only hydrogen and oxygen during operation, and is free of irritating fumes. A life of 10 years can be expected.

### Engine starting batteries

**6.25**  The capacity of engine starting batteries will depend on the size of engine and the number of cylinders.

**6.26**  Typical battery sizes are given in the table:

| Generator kW rating | Ah capacity of 24V battery |
| --- | --- |
| up to 20 | 65 – 75 |
| 20 – 40 | 75 – 85 |
| 40 – 100 | 85 – 100 |
| 150 – 200 | 110 – 140 |
| 200 – 300 | 190 – 250 |

*Battery capacities are applicable for 1500 rev/min engines with natural aspiration. For 1000 rev/min engines of the same output rating the battery capacity will be approximately 50% larger. A starting battery for use in health care and personal social services premises should be capable of turning an engine over at constant speed for 60 seconds continuously in an ambient temperature of 0°C.*

**6.27**  Starter batteries should be located as near to the engine as possible to minimise the volt drop in the connecting cable.

### Battery rooms

**6.28**  Batteries or chargers should not be located in areas of vibration or with free access by personnel.

**6.29**  Emergency batteries for the supply of slave escape luminaires are usually placed on open racks located in a separate room. Mechanical ventilation should be provided for open type batteries for removal of gases or acid fumes. Materials for lead-acid battery ventilation systems must be acid fume resistant.

**6.30**  All battery rooms should display a clear and legible notice advising staff of the type of electrolyte in the batteries and the dangers of gas explosion. Mixed storage of acid and alkaline battery electrolytes is forbidden.

**6.31**  Small battery and charger units are generally enclosed within a ventilated cubicle. For large output central installations the batteries should be located in a room separate from the UPS and lighting control equipment.

**6.32**  Recombination batteries for maintaining a UPS to data equipment, laboratory or life-support equipment may be placed next to the equipment, but must be positioned in a strong ventilated containment.

**6.33**  The sealed recombination battery does not require a purpose-built room or ventilation system to remove gases or acid fumes from the room environment when recharging. Natural ventilation should be sufficient in most locations.

# Appendix

## Combined heat and power

### General

**1.** The object of combined heat and power (CHP) is to obtain greater utilisation of the total heat content of the fuel used to drive the engine.

**2.** The overall thermal efficiency of an engine-driven generator set is not expected to be greater than 30%. This loss of 70% is represented by:

    a. friction losses of the engine revolving and sliding parts;

    b. radiation heat losses from the engine casing;

    c. exhaust gas heat and moisture losses;

    d. cylinder heat to jacket cooling water and lubricating oil;

    e. generator bearing friction losses;

    f. losses from heat generated in the copper conductors and magnetic iron circuitry;

    g. cooling fan blade dynamic friction losses.

**3.** Generator losses may amount to 5% of the total heat content of the fuel used.

**4.** The recoverable heat available to a CHP installation may amount to approximately 90% in electrical and heat unit terms.

### Heat recovery

**5.** Heat is generally recoverable at two temperature levels from the engine in heat exchangers:

    a. from the exhaust gas and moisture at the elevated temperatures between 350°C and 450°C;

    b. from the cooling water circulating through the engine water jacket and from the engine lubricating oil in intimate contact with the metal revolving and sliding parts of the engine, at temperatures between 70°C and 90°C.

**6.** For temperature and heat content values to be at the maximum for the prime mover, it is essential that continuous running at full load of the prime mover is attained. It follows that the generator providing the prime mover brake load should run continuously or for very long periods. To achieve this continuous running, additional plant may be required to ensure essential emergency supply.

**7.** The overall efficiency of electrical generation by the prime mover will determine the ultimate heat/power ratio of recovery of the rejected prime mover heat compared with electrical power generated.

**8.** The effective transfer efficiency of any heat exchanger facility will reduce the maximum value of any heat/power ratio obtained. A gas or diesel engine with an overall electrical efficiency of 30% gives an approximate heat/power ratio of 2:1. A gas turbine operates at 25% efficiency with a ratio of 3:1.

## Load optimisation

**9.** In overall terms, a health care and personal social services premises' base electrical demand represents between 10%-20% of the total energy requirement.

**10.** For a CHP set to operate continuously at full load, a suitable balance between one of the following two extreme criteria must be chosen:

- **Criterion 1** – to generate all the health care and personal social services premises' electrical demand:- For long periods during each day, the premises' heat demand cannot be supplied by the CHP set and, therefore, additional heat generation plant must be available. There would be little export of electrical power.

- **Criterion 2** – to provide all the health care and personal social services premises' heat demand:- For long periods during each day, the continuous heat available from the CHP unit will not be required. Excess heat would have to be reduced or dumped to follow the fluctuation in heat demand. There would always be an excess of electrical generation for export to the regional electricity company, subject to the terms of tariff being acceptable and the regional electricity company's electrical demand.

**11.** The balance point most economic for a particular building's requirements will be influenced by the base level of heat demand, for example domestic hot water, and seasonal heat requirements, topped up by auxiliary boiler plant. The shortfall demand in electrical power would be supplied by the regional electricity company.

**12.** The selection of several small CHP sets to run in parallel, as required by the health care and personal social services premises' demand for electrical power or heat, may be more suitable to give a better load factor. The price per kilowatt installed may be prohibitive for several small units in a single installation. This must be compared to the less efficient operation and load factors of two or three large sets, forced to operate at less than optimum running conditions.

**13.** The flow of electrical power from either the generator or the regional electricity company's supply system introduces the same generator operational requirements and direct financial tariff agreements that exist for peak lopping.

## Mechanical considerations

**14.** The generation of exhaust gas-low pressure steam reduces or eliminates the need for the generation of steam in purpose-made steam boilers, fired by light fuel oil, coal or natural gas. Steam boilers will need to be available in the event of loss of CHP generation for emergency heating of the health care and personal social services' premises.

**15.** The production of the low temperature, for example 70°C, secondary cooling water by jacket cooling water and oil heat exchange recovery can be utilised for domestic heating or as pre-heated water for admission as feed water into the high temperature heat exchangers.

**16.** In large sophisticated systems the exhaust gases from engines or gas turbines can be used in gas, oil or coal fired steam boilers as pre-heated primary inlet air in the combustion of the boiler plant fuel. Reduction in a boiler's primary air supply and self-heating requirement enhances boiler efficiency by increased steam flow, pressure and temperature at the boiler stop valve.

## Electrical considerations

**17.** The generator sets required for CHP will be required to operate on extended full load of at least 50% load factor to break even and up to 80% load factor to be economically successful.

**18.** Two generators barely satisfy the requirement for optimum operating conditions. Three generators, each of 50% of the health care and personal social services premises' base heat requirement criterion 2, would probably be required to ensure reliable service for heat and power. This allows contingencies for the regular annual engine overhauls that would be required during the summer months and to ensure an essential service supply in the event of a blackout in the regional electricity company's supply.

**19.** When three-phase, 415V AC CHP generating sets are operating in parallel with the regional electricity company supply, the health care and personal social services premises' 415V switchboard should have arrangements as shown typically in Figures 12 and 14 (see pages 27 and 29).

**20.** Figure 12 shows an arrangement with tie connections between the 415V main switchboard and the AC emergency generator 415V switchboard. The 415V essential services switchboard is connected by circuit breaker to the 415V main switchboard or the AC emergency generator 415V switchboard.

**21.** Loss of the regional electricity company supply either short-term, through a remote auto reclose switching operation, or total blackout of the grid supply, would result in the CHP generators being grossly overloaded and shut-down by operation of the electrical protection. The provision of high speed main protection to detect the loss of normal supply is recommended to open the circuit breaker at the supply transformer. Remotely controlled synchronising facilities should be provided at the LV circuit breakers to permit later synchronisation between supplies from the generator control panels. The type of main protection is discussed in HTM2011, Emergency electrical services, 'Operational management', paragraph 3.5.

**22.** In the arrangement in Figure 12 on page 27, CHP generator sets operate in parallel with the regional electricity company supply. A total loss of either the CHP generating capacity or the regional electricity company supply will result in a smooth changeover to either of the electrical supplies at the generator-essential tie circuit breaker or the transformer HV circuit breaker, respectively. If the fault to the normal supply occurs at the 415V main switchboard, the result will be an automatic opening of the main-essential tie circuit breaker, resulting in electrical demand reducing to only the essential services requirement and a loss of a major part of the heat generation to the health care and personal social services premises.

**23.** During times of severe local electrical storm it is recommended as a precaution that a health care and personal social services premises CHP electrical system operates separately from the regional electricity company normal supply where HV overhead transmission lines are used for the incoming supply.

**24.** If the CHP load demand requires AC generators over 500kVA rating, generation at the three-phase 11kV busbars should be considered depending on current and disconnecting protection. Electrical power generation at 11kV would give better operating conditions and dispense with the need for external on-load voltage control. A three-phase, 415V AC emergency generator back-up system should also be available to ensure essential emergency supply in the event of loss of the three-phase 11kV/433V supply transformer, as shown in Figure 14 on page 29.

**25.** 11kV on-load voltage regulation by double induction regulator control may be required for CHP installations operating at LV and exporting bulk electrical power to the regional electricity company. This may be necessary to maintain stability in generators and approved voltage levels for the two systems under extreme operating conditions.

**26.** On-load tap changing equipment for small transformers below 10MVA rating is unusual. The cost of the on-load tap changer equipment can equal the cost of the transformer, and the tap changer can be physically equal in size to the transformer.

**27.** Induction generators operate at a leading power factor to the incoming normal supply. This is mirror imaged by the lagging power factor of the incoming supply transformer.

**28.** When induction generators are operating in parallel with the normal supply, a loss of normal supply without sufficient health care and personal social services premises synchronous generating capacity, will entail a collapse of health care and personal social services premises generation. In such situations the health care and personal social services premises generation must be designed to stand alone when isolated from the normal supply.

**29.** It is recommended that synchronous AC emergency generating sets should always be available to provide essential services supply requirements where CHP induction generator sets are provided.

**30.** The equivalent maximum full load inductive reactive power generated by the health care and personal social services premises' synchronous generators must be sufficient to maintain the induction generator magnetisation and health care and personal social services premises' inductive load demand, and hence system voltage and dynamic stability. For equally rated induction generators and AC synchronous generators, the reactive power required for magnetisation of the stator cores is equally rated and equivalent to the short circuit ratio of each generator. On this estimate, for parallel operation it will require an AC synchronous generator of at least twice the reactive rating of an induction generator to provide the minimum necessary induction generator core magnetisation, and at the same time maintain the system voltage and AC generator stability, at best, unity power factor.

**31.** An induction generator without low voltage frequency protection for overall shutdown may suffer sudden removal of generator load owing to the loss of magnetisation current provided from the supply, that is, loss of normal supply. In this case the overspeed trip protection of the engine may occur.

**32.** Induction generators do not require the synchronising facilities required for synchronous generators, and thus the electrical installation cost will be lower. The risk of over-simplification of the installation must be carefully balanced, that is, the economic advantage offered against security of emergency supply.

# Other publications in this series

(Given below are details of all Health Technical Memoranda available from HMSO. HTMs marked (*) are currently being revised, those marked (†) are out of print. Some HTMs in preparation at the time of publication of this HTM are also listed.)

1  Anti-static precautions: rubber, plastics and fabrics*†
2  Anti-static precautions: flooring in anaesthetising areas (and data processing rooms)*, 1977.
3  –
4  –
5  Steam boiler plant instrumentation†
6  Protection of condensate systems: filming amines†
7  Electrical services: supply and distribution*†
8  –
9  –
10  Sterilizers*†
12  –
13  –
14  Abatement of electrical interference*†
15  Patient/nurse call systems†
16  –
17  Health building engineering installations: commissioning and associated activities, 1978.
18  Facsimile telegraphy: possible applications in DGHs†
19  Facsimile telegraphy: the transmission of pathology reports within a hospital – a case study†
20  Electrical safety code for low voltage systems (in preparation)
21  Electrical safety code for high voltage systems*†
22  Piped medical gases, medical compressed air and medical vacuum installations*†
22  Supp. Permit to work system: for piped medical gases etc†
23  Access and accommodation for engineering services†
24  –
25  –
26  Commissioning of oil, gas and dual fired boilers: with notes on design, operation and maintenance†
27  Cold water supply storage and mains distribution* [Revised version will deal with water storage and distribution], 1978.
28 to 53  –

## Component Data Base

54  User manual, 1989.
55  Windows, 1989.
56  Partitions, 1989.
57  Internal glazing, 1989.
58  Internal doorsets, 1989.
59  Ironmongery, 1989.
60  Ceilings, 1989.
61  Flooring, 1989.

62  Demountable storage systems, 1989.
63  Fitted storage systems, 1989.
64  Sanitary assemblies, 1989.
65  Signs†
66  Cubicle curtain track, 1989.
67  Laboratory fitting-out (in preparation)
68  Ducts and panels for sanitary assemblies (in preparation)
69  Protection (in preparation)
70  Fixings (in preparation)
71 to 80  –

## Firecode

81  Firecode: fire precautions in new hospitals, 1987.
82  Firecode: alarm and detection systems, 1989.
83  Fire safety in health care premises: general fire precautions*†
84  Fire precautions in new residential care premises (in preparation)
85  [Revision to Home Office draft guidance in preparation]
86  Firecode: assessing fire risks in existing hospital wards, 1987
87  Firecode: textiles and furniture, 1987.
88  Fire safety in health care premises: guide to fire precautions in NHS housing in the community for mentally handicapped/ill people, 1986.

## New HTMs in preparation

Lifts
Mains signalling
Legionnaires Disease
Telecommunications
Washers for sterile production
Special ventilation systems
Risk management and quality assurance

Health Technical Memoranda published by HMSO can be purchased from HMSO Bookshops in London (post orders to PO Box 276, SW8 5DT), Edinburgh, Belfast, Manchester, Birmingham and Bristol or through good booksellers. HMSO provide a copy service for publications which are out of print; and a standing order service.

Enquiries about Health Technical Memoranda (but not orders) should be addressed to:
NHS Estates, Department of Health, Room 313,
Euston Tower, 286 Euston Road, London NW1 3DN.

# About NHS Estates

NHS Estates is an Executive Agency of the Department of Health and is involved with all aspects of health estate management, development and maintenance. The Agency has a dynamic fund of knowledge which it has acquired during 30 years of working in the field. Using this knowledge NHS Estates has developed products which are unique in range and depth. These are described below.

NHS Estates also makes its experience available to the field through its consultancy services.

Enquiries should be addressed to: NHS Estates, Euston Tower, 286 Euston Road, London NW1 3DN. Telephone: 071-388 1188.

## Some other NHS Estates products

**Activity DataBase** – a computerised system for defining the activities which have to be accommodated in spaces within health buildings. *NHS Estates*

**Design Guides** – complementary to Health Building Notes, Design Guides provide advice for planners and designers about subjects not appropriate to the Health Building Notes series. *HMSO*

**Estatecode** – user manual for managing a health estate. Includes a recommended methodology for property appraisal and provides a basis for integration of the estate into corporate business planning. *HMSO*

**Capricode** – a framework for the efficient management of capital projects from inception to completion. *HMSO*

**Concode** – outlines proven methods of selecting contracts and commissioning consultants. Both parts reflect official policy on contract procedures. *NHS Estates*

**Works Information Management System** – a computerised information system for estate management tasks, enabling tangible assets to be put into the context of servicing requirements. *NHS Estates*

**Option Appraisal Guide** – advice during the early stages of evaluating a proposed capital building scheme. Supplementary guidance to Capricode. *HMSO*

**Health Building Notes** – advice for project teams procuring new buildings and adapting or extending existing buildings. *HMSO*

**Health Guidance Notes** – an occasional series of publications which respond to changes in Department of Health policy or reflect changing NHS operational management. Each deals with a specific topic and is complementary to a related Health Technical Memorandum. *HMSO*

**Encode** – shows how to plan and implement a policy of energy efficiency in a building. *HMSO*

**Firecode** – for policy, technical guidance and specialist aspects of fire precautions. *HMSO*

**Nucleus** – standardised briefing and planning system combining appropriate standards of clinical care and service with maximum economy in capital and running costs. *NHS Estates*

**Concise** – software support for managing the capital programme. Compatible with Capricode. *NHS Estates*

Items noted "HMSO" can be purchased from HMSO Bookshops in London (post orders to PO Box 276, SW8 5DT), Edinburgh, Belfast, Manchester, Birmingham and Bristol or through good booksellers. Details of their standing order service are given at the front of this publication.

Enquiries about NHS Estates products should be addressed to: NHS Estates, The Publications Unit, Department of Health, Room 540, Euston Tower, 286 Euston Road, London NW1 3DN.

## NHS Estates consultancy service

Designed to meet a range of needs from advice on the oversight of estates management functions to a much fuller collaboration for particularly innovative or exemplary projects.

Enquiries should be addressed to: NHS Estates, Room 335 (address as above).

Printed in the United Kingdom for HMSO.
Dd 295512   C15   10/92   17647